Image Sensors - Digital Imaging Systems and Applications

Edited by Francisco Javier Gallegos-Funes

Published in London, United Kingdom

Image Sensors - Digital Imaging Systems and Applications
http://dx.doi.org/10.5772/intechopen.1005691
Edited by Francisco Javier Gallegos-Funes

Contributors
Aline Nunes de Souza, Anderson Felipe Weschenfelder, Armando Leopoldo Keller, Chunling Tu, Davi Schmitz, Etienne A. van Wyk, Fadhil Hidayat, Figo Agil Alunjati, Jean Schmith, Jean Schmith, Jean Schmith, Jéssica Diehl, Kenji Hara, Kohei Inoue, Pius A. Owolawi, Renan Santos dos Santos, Rodrigo Marques de Figueiredo, Rodrigo Marques de Figueiredo, Rodrigo Marques de Figueiredo, Rotimi-Williams Bello, Ulva Elviani, Vitor Camargo Nardelli, Vitor Camargo Nardelli, Vitor Camargo Nardelli

First published in London, United Kingdom, 2025 by IntechOpen
IntechOpen is the global imprint of INTECHOPEN LIMITED, registered in England and Wales, registration number: 11086078, 167-169 Great Portland Street, London, W1W 5PF, United Kingdom

For EU product safety concerns: IN TECH d.o.o., Prolaz Marije Krucifikse Kozulić 3, 51000 Rijeka, Croatia, info@intechopen.com or visit our website at intechopen.com.

British Library Cataloguing-in-Publication Data
A catalogue record for this book is available from the British Library

Image Sensors - Digital Imaging Systems and Applications
Edited by Francisco Javier Gallegos-Funes
p. cm.
Print ISBN 978-1-83634-968-6
Online ISBN 978-1-83634-967-9
eBook (PDF) ISBN 978-1-83634-969-3

Meet the editor

Francisco J. Gallegos-Funes received his Ph.D. in communications and electronics from the Instituto Politécnico Nacional de México (National Polytechnic Institute of Mexico) in 2003. He is currently an associate professor in the Escuela Superior de Ingeniería Mecánica y Eléctrica (Mechanical and Electrical Engineering Higher School) at the same institute. His areas of scientific interest include signal and image processing, filtering, steganography, segmentation, pattern recognition, biomedical signal processing, sensors, and real-time applications.

Contents

Preface

Research on digital imaging systems and computer vision systems has been increasing in the last few years. An image sensor is one of the main blocks in a digital imaging system, such as a digital still or video camera. This sensor captures a scene for use in various imaging applications, including machine vision, time-of-flight (TOF) imaging, topographic imaging, three-dimensional high-definition television (3D-HDTV), and optical molecular imaging systems. For these reasons, the capabilities of image sensors offer the answer to many researchers who utilize digital images.

This book contains a select number of chapters based on digital imaging systems, the result of research conducted by several researchers and professionals who have made significant contributions to the field of digital imaging systems. This book includes the following chapters: Chapter 1, *360-Degree Image Denoising Using Grid-Based Image Decomposition*, proposes image denoising by dividing a 360-degree image into multi-decomposed images using a superimposed grid of six-element grids covering a spherical surface. The resulting rectangular images are denoised using the non-local means (NLM) filter.

Chapter 2, *Perspective Chapter: Advancement in CCTV Video Analytics – Leveraging Face Recognition for Enhanced Security Solutions*, explores the latest advancements in CCTV video analytics and its applications, specifically in the domain of facial recognition.

Chapter 3, *Analysis of Similarity Structures for Star Identification in Blurred Images*, proposes a methodology for detecting stars in blurred images by developing a mathematical model based on probabilistic approaches, exploring the relationship between the light dispersion of stars and a Gaussian distribution.

Chapter 4, *Fire Detection Using Image Processing and Machine Learning*, proposes a two-stage fire detection method in agribusiness contexts utilizing k-Nearest Neighbors (kNN) to classify images that contain fire.

Chapter 5, *Object Detection Algorithms for Digital Imaging Applications: A Review* reviews the evolution of 2D object detection algorithms for digital imaging applications, focusing on their developments, models, applications, datasets, evaluation metrics, strengths and weaknesses for a better understanding of their landmarks and contributions to the advancement of the field.

Chapter 6, *Usage of Wavelets in Image-Based Steganography*, enables a comprehensive comparison of the steganographic behavior of models that utilize low-frequency, high-frequency, and multi-level sub-bands, assessing the impact of various wavelet families, scaling factors, and decomposition levels.

I hope that readers find these chapters interesting and informative and that they serve as a basis for future research or projects.

Francisco Javier Gallegos-Funes
Escuela Superior de Ingeniería Mecánica y Eléctrica,
Instituto Politécnico Nacional,
Mexico City, Mexico

Chapter 1

360-Degree Image Denoising Using Grid-Based Image Decomposition

Kenji Hara and Kohei Inoue

Abstract

Equirectangular projection images are widely used as input for 360-degree images. However, due to the agglomeration of pixels in the polar regions, which requires special processing, existing image denoising methods designed for general planar images cannot be directly applied without modification. In this chapter, we propose denoising the images by dividing a 360-degree image into multi-decomposed images using a superimposed grid of six element grids covering a spherical surface. These resulting rectangular images, which have nearly uniform spatial resolution, are then denoised using the non-local means (NLM) method. The effectiveness of this method is demonstrated by applying it to real 360-degree images.

Keywords: 360-degree image denoising, non-local means, overset grid method, multi-decomposition of 360-degree image, upper latitude minimization

1. Introduction

Image denoising is often applied as preprocessing for a wide variety of computer vision tasks, for example, image segmentation, contrast enhancement, frequency decomposition, and feature extraction. Patch-based image denoising methods, such as non-local means (NLM) [1], block matching and 3D filtering (BM3D) [2], non-local sparse model (NLSM) [3], low-rank continuous orthogonal matching pursuit (LR-COMP) [4], patch-based CNN (PCNN) [5], and patch-based adaptive temporal filter (PATF) [6], whose weights are calculated on the basis of an image patch similarity measures, are still widely used and the considered state-of-the-art algorithms. Patch-based image denoising methods are designed for ordinary planar images, and if applied directly to spherical images, such as 360-degree images (hereinafter referred to as "360-degree images"), they fail to account for the following properties of latitude-longitude images, which are the standard representation format for 360-degree images. This oversight can lead to incorrect calculations of patch similarity and weighted averages, resulting in undesirable denoising: (1) the resolution is spatially non-uniform due to the significant variation in the area ratio covered by each pixel, and (2) exceptional processing is required near the poles, corresponding to the top and bottom areas of the latitude-longitude image.

Recently, an overset grid method called the Yin-Yang grid was independently proposed by Kageyama and Sato [7] and Purser [8] for calculating spheres and spherical shells in the field of earth science. The representation of a 360-degree image using the

Yin-Yang grid offers several advantages over the latitude-longitude image representation: (1) there are no singularities in the top and bottom regions of the image, eliminating the need for special handling near the poles; (2) the area ratio covered by each pixel is approximately 0.71 or greater, resulting in minimal variation and a more uniform spatial resolution than a latitude-longitude image; and (3) the number of pixels is approximately 75% of that in a latitude-longitude image, reducing the number of redundant pixels. Focusing on these advantages of the Yin-Yang overset grid, several applications in the computer vision have been reported including feature point detection [9], saliency map generation [10], and superpixel segmentation [11] for 360-degree images.

We propose a patch-based framework for denoising 360-degree images by using a new overset grid method that improves the resolution uniformity of the Yin-Yang grid. To achieve better spatial resolution uniformity, we formulate an overset grid method that decomposes the sphere into a six-element grid based on upper latitude minimization. Next, image denoising is performed using standard NLM filter on the decomposed images that have been obtained with uniform spatial resolution. Finally, these denoised images are merged to reconstruct the 360-degree image, achieving highly accurate 360-degree image denoising.

2. Yin-Yang grid

In the decomposition of a 360-degree image using the Yin-Yang overset grid, the latitude-longitude image is split into two low-latitude images corresponding to the Yin and Yang overset grids. The Yin grid has a latitude range of $-45^{\circ} \sim 45^{\circ}$ and a longitude range of $-135^{\circ} \sim 135^{\circ}$. The Yang grid is a three-dimensional rotation of the Yin grid (see **Figure 1**). The Yin-Yang grid offers the following advantages over the usual latitude-longitude grid.

1. No special processing near the poles is required because there are no singularities at the poles.

2. Since the area ratio of pixels corresponding to grid points is about 0.71 or higher, the variation in lattice size is minimal, resulting in a nearly uniform spatial resolution across the grid.

3. Since both the Yin and Yang overset grids are based on latitude-longitude lattices, the Yin-Yang overset grid can be directly displayed, allowing the scene to remain easily interpretable.

4. The number of grid points is about 75% of that in a standard latitude-longitude grid, making it less computationally intensive.

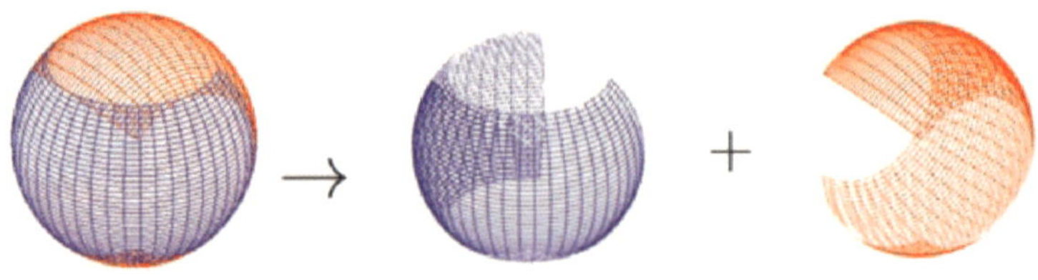

Figure 1.
Yin-Yang grid method.

3. Proposed method

3.1 Multi-decomposition of 360-degree images based on extention of Yin-Yang grid

We present a multi-decomposition method for 360-degree images based on latitude upper bound minimization to improve the uniformity of spatial resolution in decomposed 360-degree images using the Yin-Yang overset grid geometry. We fix the Cartesian coordinate system $O - xyz$ in three-dimensional Euclidean space and assume that the x-axis directions (and similarly, the z-axis and y-axis directions) are from the origin O toward latitude 0 and longitude 0. Assume that the y-axis directions (and similarly, the x-axis and z-axis directions) are oriented from the origin O toward latitude 0 and longitude $\pi/2$. Additionally, assume that the z-axis directions (and similarly, the y-axis and x-axis directions) are oriented from the origin O toward the North Pole. In this case, latitude (positive for north latitude and negative for south latitude) is denoted as ϕ (with ϕ' and ϕ'' for corresponding values), and longitude is denoted as λ (with λ' and λ'' for corresponding values). Using these latitude and longitude coordinates, the coordinates (x, y, z) of a point on the unit sphere centered at the origin O are expressed as follows.

$$x = \cos\phi\cos\lambda = \cos\phi'\sin\lambda' = \sin\phi'' \tag{1}$$

$$y = \cos\phi\sin\lambda = \sin\phi' = \cos\phi''\cos\lambda'' \tag{2}$$

$$z = \sin\phi = \cos\phi'\cos\lambda' = \cos\phi''\sin\lambda'' \tag{3}$$

The latitude-longitude coordinate systems corresponding to Eqs. (1)–(3) are called Σ, Σ', and Σ'', respectively. The Yin-Yang grid can then be regarded as a spherical partition that allows overlap, using the latitude-longitude coordinate systems Σ and Σ'. This configuration ensures for each point on the unit sphere, the point is represented as close as possible to zero (i.e., at low latitude) in either of the two latitudes ϕ and ϕ'' (see **Figure 1**). The Yin-Yang grid allows the decomposition of a 360-degree image into the two images, where every pixel has a latitude of $\pi/4 = 45^\circ$ or less, and the spatial resolution remains nearly uniform. In contrast, we now describe a method for multi-decomposing a 360-degree image into multiple images with even more uniform spatial resolution than that achieved using the Yin-Yang overset grid. This is achieved by utilizing all three latitude-longitude coordinate systems Σ, Σ', and Σ''. For eight points whose Cartesian coordinates on the unit sphere are $(x, y, z) = \left(\pm\sqrt{1/3}, \pm\sqrt{1/3}, \pm\sqrt{1/3}\right)$ (with the signs being complex sign optional), the magnitudes of the latitudes in the latitude-longitude coordinate systems Σ, Σ', and Σ'' are equal to each other. Therefore, in the latitude-longitude coordinate system Σ, the area below this latitude value is represented as below.

$$-\mathrm{Arcsin}\frac{1}{\sqrt{3}} \leq \phi \leq \mathrm{Arcsin}\frac{1}{\sqrt{3}}, \tag{4}$$

where $\mathrm{Arcsin}(y)$ is an inverse function of $y = \sin x \ \ (x \in [-\pi/2, \pi/2])$.

The low-latitude region (4), rotated 90° around the x axis, can be expressed under the latitude-longitude coordinate system Σ as follows.

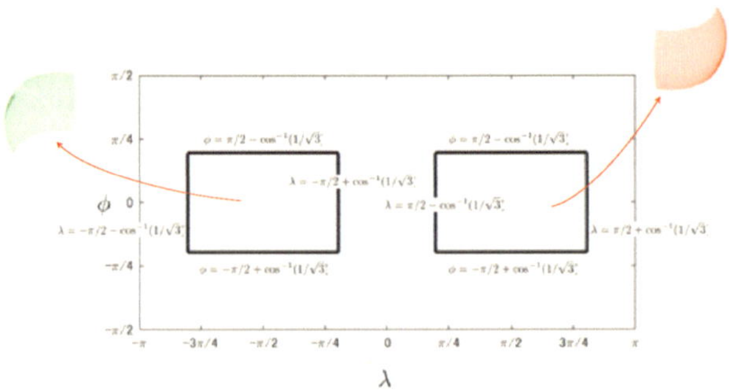

Figure 2.
Division and extraction of regions on a spherical surface.

$$
\begin{gathered}
-\pi \leq \lambda \leq \sin\left(\frac{1}{\sqrt{3}\cos\phi}\right) - \pi \\
\text{or} \\
-\sin\left(\frac{1}{\sqrt{3}\cos\phi}\right) \leq \lambda \leq \sin\left(\frac{1}{\sqrt{3}\cos\phi}\right) \\
\text{or} \\
\pi - \sin\left(\frac{1}{\sqrt{3}\cos\phi}\right) \leq \lambda < \pi.
\end{gathered}
\tag{5}
$$

The polar point in the latitude-longitude coordinate system Σ of the boundary line of the area on the sphere (5) is given by $(\lambda, \phi) = \left(\mathrm{Arcsin}(1/\sqrt{3}) - \pi, 0\right)$, $\left(-\mathrm{Arcsin}(1/\sqrt{3}), 0\right)$, $\left(\mathrm{Arcsin}(1/\sqrt{3}), 0\right)$, $\left(\pi - \mathrm{Arcsin}(1/\sqrt{3}), 0\right)$. Therefore, we divide and extract two regions on the sphere in the latitude-longitude coordinate system Σ, as expressed by the following equation (see **Figure 2**).

$$
\begin{gathered}
-\frac{\pi}{2} - \mathrm{Arccos}\left(\frac{1}{\sqrt{3}}\right) \leq \lambda \leq -\frac{\pi}{2} + \mathrm{Arccos}\left(\frac{1}{\sqrt{3}}\right), \quad -\frac{\pi}{2} + \mathrm{Arccos}\left(\frac{1}{\sqrt{3}}\right) \leq \phi \leq \frac{\pi}{2} - \mathrm{Arccos}\left(\frac{1}{\sqrt{3}}\right) \\
\text{or} \\
\frac{\pi}{2} - \mathrm{Arccos}\left(\frac{1}{\sqrt{3}}\right) \leq \lambda \leq \frac{\pi}{2} + \mathrm{Arccos}\left(\frac{1}{\sqrt{3}}\right), \quad -\frac{\pi}{2} + \mathrm{Arccos}\left(\frac{1}{\sqrt{3}}\right) \leq \phi \leq \frac{\pi}{2} - \mathrm{Arccos}\left(\frac{1}{\sqrt{3}}\right),
\end{gathered}
\tag{6}
$$

where $\mathrm{Arccos}(y)$ is an inverse function of $y = \cos x \ (x \in [0, \pi])$.

Next, these regions on the sphere (6) are first rotated 90° around the x-axis and then 90° around the y-axis, respectively. The resulting regions on the sphere are expressed by the equations below in the latitude-longitude coordinate system Σ (see **Figure 3**).

$$
\begin{gathered}
-\pi \leq \lambda \leq \mathrm{Arcsin}\left(\frac{1}{\sqrt{3}\cos\phi}\right) - \pi, \quad \mathrm{Arctan}\left(\sqrt{2}\cos\lambda\right) \leq \phi \leq -\mathrm{Arctan}\left(\sqrt{2}\cos\lambda\right) \\
\text{or} \\
-\mathrm{Arcsin}\left(\frac{1}{\sqrt{3}\cos\phi}\right) \leq \lambda \leq \mathrm{Arcsin}\left(\frac{1}{\sqrt{3}\cos\phi}\right), \quad -\mathrm{Arctan}\left(\sqrt{2}\cos\lambda\right) \leq \phi \leq \mathrm{Arctan}\left(\sqrt{2}\cos\lambda\right) \\
\text{or} \\
\pi - \mathrm{Arcsin}\left(\frac{1}{\sqrt{3}\cos\phi}\right) \leq \lambda \leq \pi, \quad \mathrm{Arctan}\left(\sqrt{2}\cos\lambda\right) \leq \phi \leq -\mathrm{Arctan}\left(\sqrt{2}\cos\lambda\right),
\end{gathered}
\tag{7}
$$

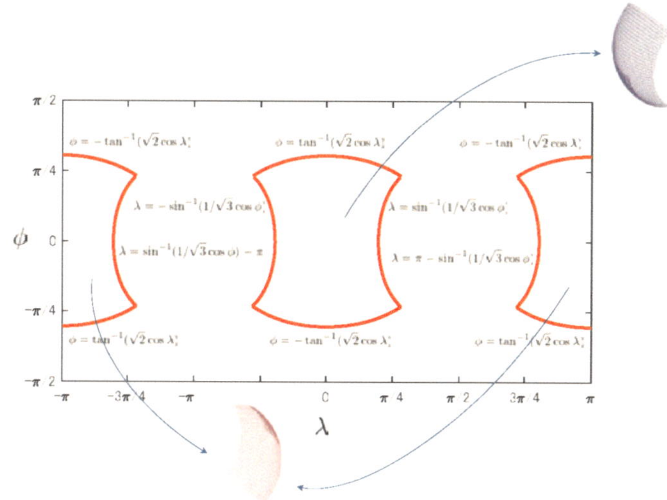

Figure 3.
Division and extraction of regions on a spherical surface.

where Arctan(y) is an inverse function of $y = \tan x \ (x \in [-\pi/2, \pi/2])$. Note that while Eq. (7) is divided into three regions in the latitude-longitude coordinate system Σ, Eq. (7) is connected into a single region on the sphere. Therefore, Eq. (7) represents two regions on the sphere. Now, rotate the on-sphere region (6) simultaneously 90° around the z-axis and then 90° around the y-axis. The area on the sphere obtained from this rotation is expressed in the latitude-longitude coordinate system Σ as follows (note that $\pm$ is assumed to follow the same order of for both signs. See **Figure 4**).

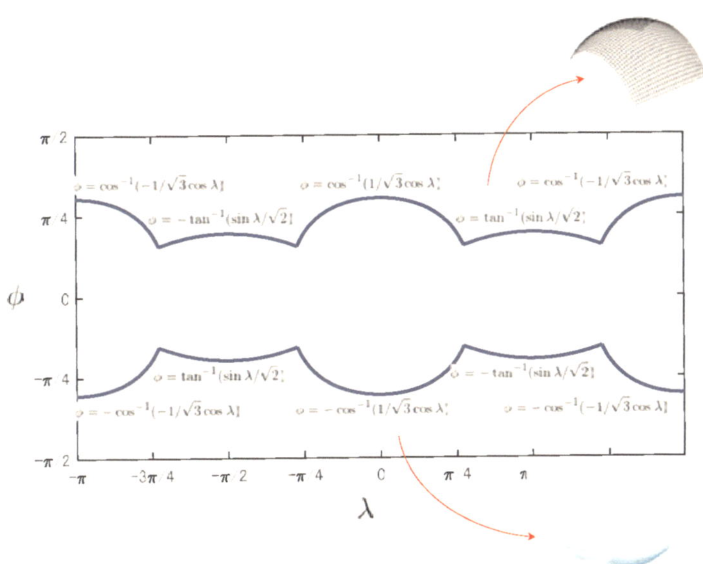

Figure 4.
Division and extraction of regions on a spherical surface.

$$
\begin{gathered}
-\pi \le \lambda \le -\pi + \mathrm{Arccos}\sqrt{\frac{3}{7}},\ \ \mathrm{Arccos}\left(-\frac{1}{\sqrt{3}\cos\lambda}\right) \le \pm\phi \le \frac{\pi}{2} \\
\text{or} \\
-\pi + \mathrm{Arccos}\sqrt{\frac{3}{7}} \le \lambda \le -\mathrm{Arccos}\sqrt{\frac{3}{7}},\ \ -\mathrm{Arctan}\left(\frac{\sin\lambda}{\sqrt{2}}\right) \le \pm\phi \le \frac{\pi}{2} \\
\text{or} \\
-\mathrm{Arccos}\sqrt{\frac{3}{7}} \le \lambda \le \mathrm{Arccos}\sqrt{\frac{3}{7}},\ \ -\mathrm{Arccos}\left(\frac{1}{\sqrt{3}\cos\lambda}\right) \le \pm\phi \le \frac{\pi}{2} \\
\text{or} \\
\mathrm{Arccos}\sqrt{\frac{3}{7}} \le \lambda \le \pi - \mathrm{Arccos}\sqrt{\frac{3}{7}},\ \ \mathrm{Arctan}\left(\frac{\sin\lambda}{\sqrt{2}}\right) \le \pm\phi \le \frac{\pi}{2} \\
\text{or} \\
\pi - \mathrm{Arccos}\sqrt{\frac{3}{7}} \le \lambda < \pi,\ \ \mathrm{Arccos}\left(-\frac{1}{\sqrt{3}\cos\lambda}\right) \le \pm\phi \le \frac{\pi}{2}.
\end{gathered}
\tag{8}
$$

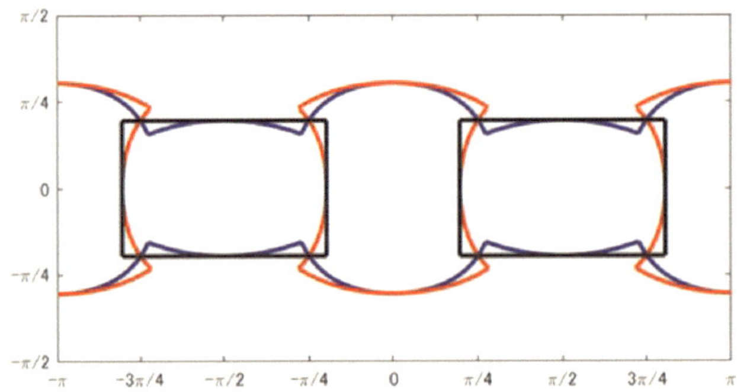

Figure 5.
*Superimposed display of the boundaries of the regions on the sphere in **Figures 2–4**.*

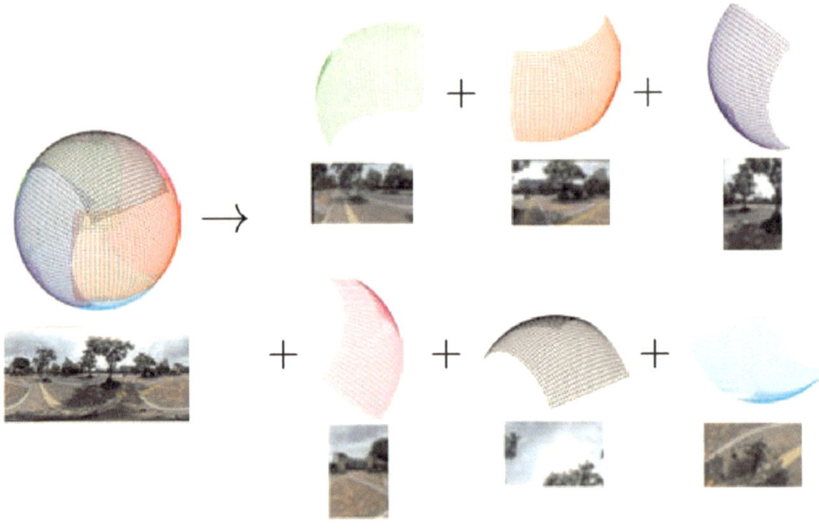

Figure 6.
Our overset grid method with latitude upper bound minimization.

Figure 5 shows the spherical regions (6)–(8) simultaneously in the latitude and longitude coordinate system Σ. The on-sphere region shown in **Figure 5** covers the entire sphere, and the upper bound of latitude, $\text{Arcsin}(1/\sqrt{3}) \approx 35.3^{\circ}$, being smaller than the upper bound of latitude, 45°, in the Yin-Yang grid. This indicates that the 360-degree image is divided into regions with more uniform spatial resolution than the Yin-Yang grid. Each of the on-sphere regions (6)–(8) consists of two connected regions. Therefore, the 360-degree image is divided into six overlapping regions. This spherical image decomposition is shown in **Figure 6**.

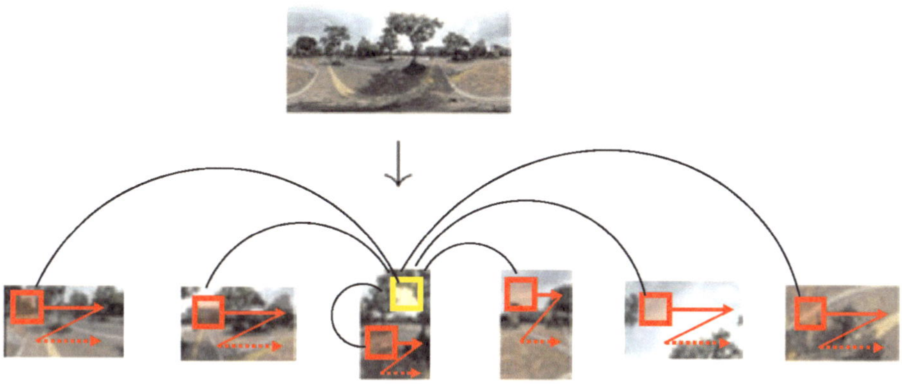

Figure 7.
Measuring the similarity between patches within a decomposed image and between patches across other decomposed images using weighted averaging.

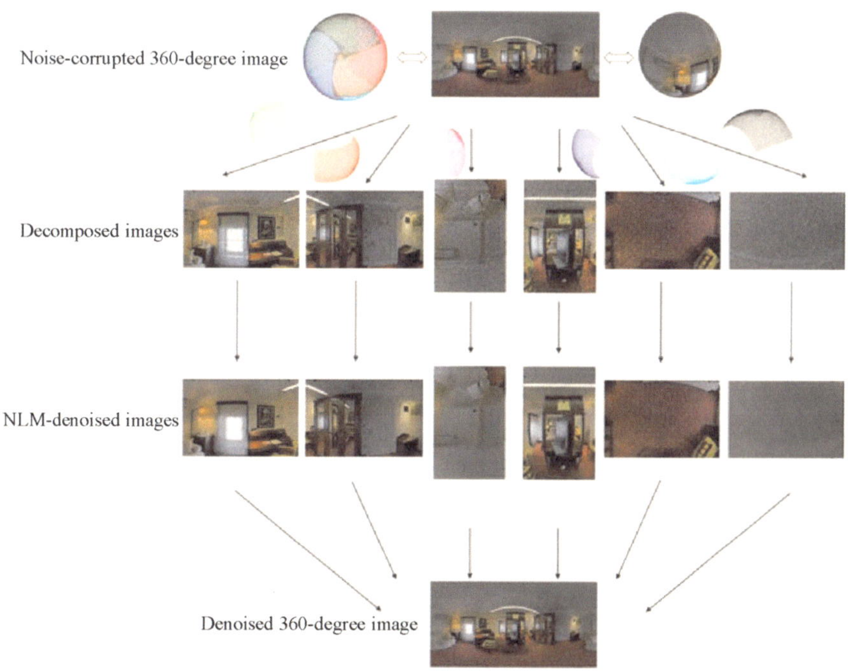

Figure 8.
Overall flow of our method.

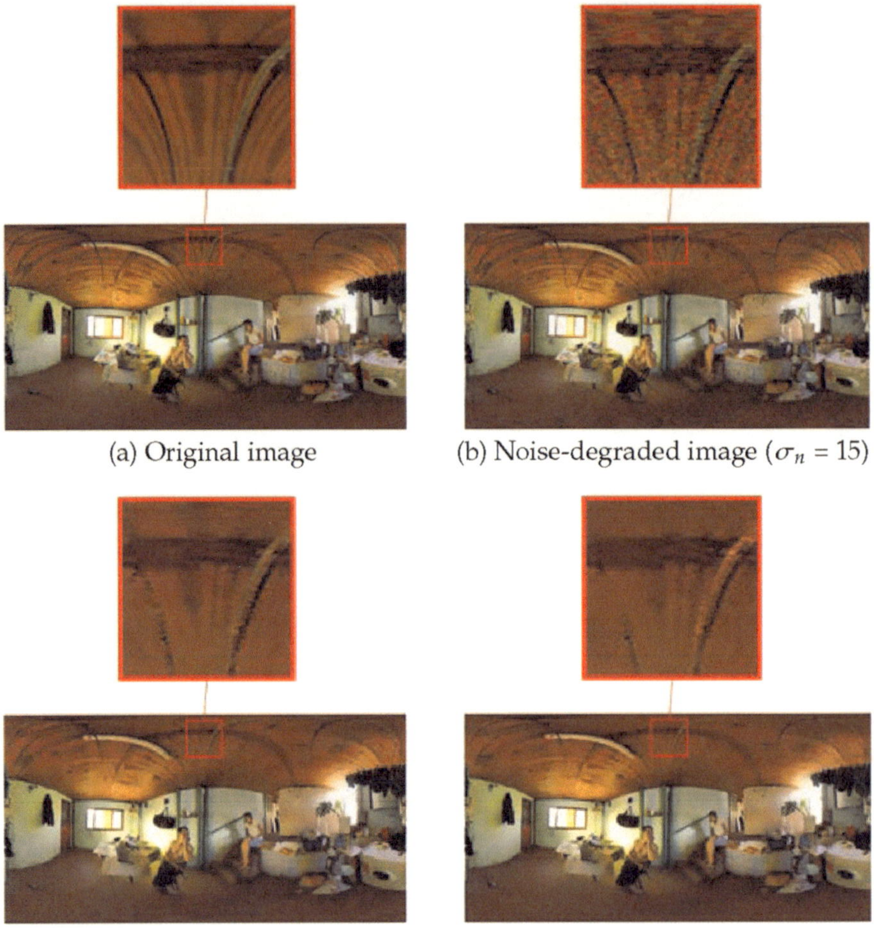

(a) Original image (b) Noise-degraded image (σ_n = 15)

(c) Baseline method (d) Our proposed method

Figure 9.
Visual comparison of 360-degree image denoising on image "2" with noise level of 15. (a) Original and enlarged images. (b) Degraded and enlarged images by Gaussian noise $N(0, 15)$ added from all directions. (c) Denoised and enlarged images with the baseline method. (d) Denoised and enlarged images with the proposed method.

3.2 Resolution uniformity of multi-decomposed images

As mentioned in Section 2.2, for a decomposed 360-degree image using the Yin-Yang overset grid, the area ratio of the smallest grid to the largest grid is approximately $1/\sqrt{2}\approx 0.71$. In contrast, the area of each grid acquired by multi-resolution decomposition of the 360-degree image reaches its maximum value of $\delta\phi\delta\lambda$ when $\phi = 0$, and the minimum value is $\delta\phi\delta\lambda\cos(\pm\mathrm{Arcsin}(1/\sqrt{3})) = \sqrt{2/3}\delta\phi\delta\lambda\approx 0.82\delta\phi\delta\lambda$ when $\phi = \pm\mathrm{Arcsin}(1/\sqrt{3})$, as described in the previous section. Therefore, the area ratio of the smallest grid to the largest grid is about 0.82, indicating that the decomposed image obtained by multi-decomposition of a 360-degree image is closer to 1 (about 0.71) than the Yin-Yang overset grid, resulting in a more spatially uniform resolution.

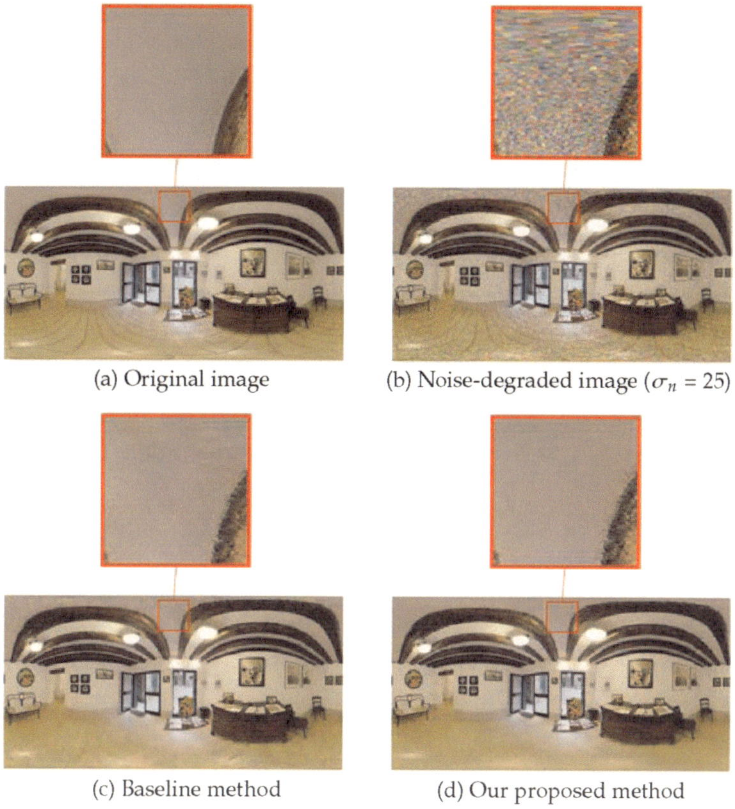

Figure 10.
Visual comparison of 360-degree image denoising on image "3" with noise level of 25. (a) Original and enlarged images. (b) Degraded and enlarged images by Gaussian noise $N(0, 25)$ added from all directions. (c) Denoised and enlarged images with the baseline method. (d) Denoised and enlarged images with the proposed method.

3.3 Number of pixels in multi-decomposed images

As described in the previous section, if the number of vertical pixels in a 360-degree latitude-longitude image is M and the number of horizontal pixels is N, the total number of pixels in the image decomposed using the Yin-Yang grid is $3MN/4 = 0.75MN$. In contrast, the grid on the sphere corresponding to each of the decomposed images obtained by multi-decomposition of the 360-degree image has a latitude range of $\phi \in \left[-\pi/2 + \text{Arccos}(1/\sqrt{3}), \pi/2 - \text{Arccos}(1/\sqrt{3})\right]$ and a longitude range of $\lambda \in \left[-\pi/2 - \text{Arccos}(1/\sqrt{3}), -\pi/2 + \text{Arccos}(1/\sqrt{3})\right]$. Consequently, the number of vertical pixels and horizontal pixels in each resolved image are $\left(1 - (2/\pi)\text{Arccos}(1/\sqrt{3})\right)M \approx 0.39M$ and $(1/\pi)\text{Arccos}(1/\sqrt{3})N \approx 0.3N$, respectively. Therefore, the number of whole pixels in the decomposed images is approximately $(6/\pi)\text{Arccos}(1/\sqrt{3})\left(1 - (2/\pi)\text{Arccos}(1/\sqrt{3})\right)MN \approx 6 \times 0.39M \times 0.3N \approx 0.7MN$. Thus, it can be observed that the multi-decomposition of the 360-degree image reduces the total number of pixels more than the Yin-Yang overset grid, as the reduction in the number of entire pixels relative to the standard latitude-longitude image is about 30% (compared to 25% with the Yin-Yang grid).

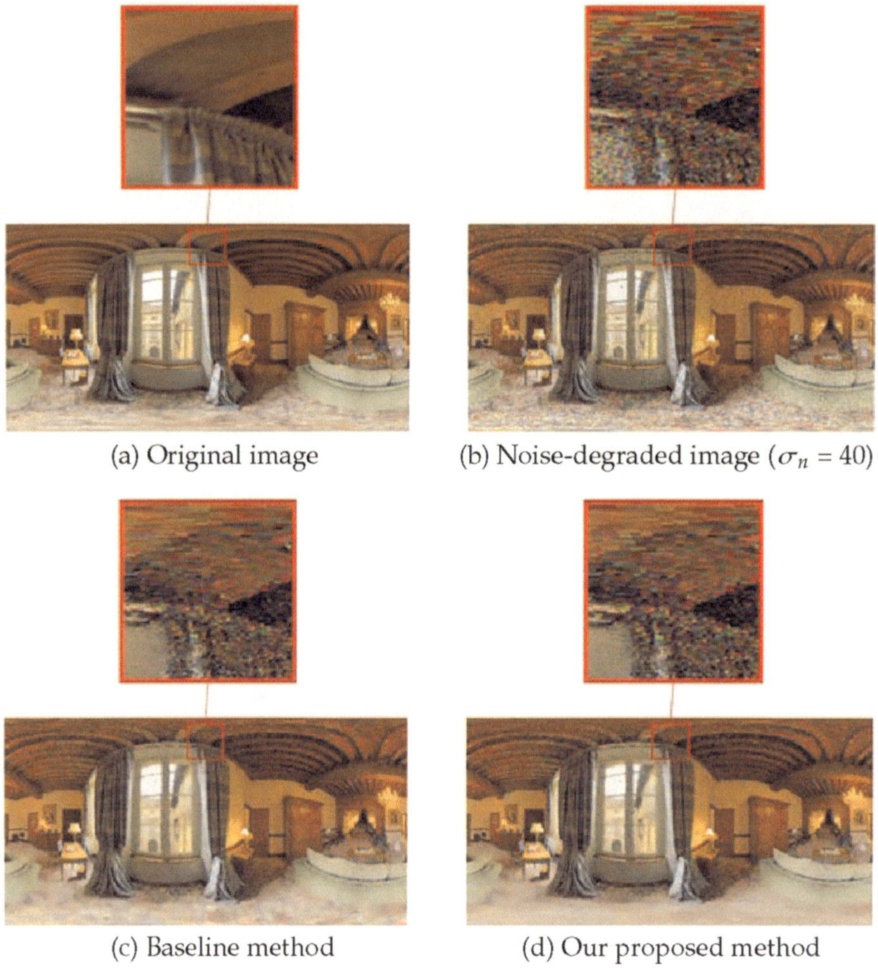

Figure 11.
Visual comparison of 360-degree image denoising on image "4" with noise level of 40. (a) Original and enlarged images. (b) Degraded and enlarged images by Gaussian noise N(0, 40) added from all directions. (c) Denoised and enlarged images with the baseline method. (d) Denoised and enlarged images with the proposed method.

3.4 Pach-based denoising and integration of multi-decomposed images

The multi-decomposed images with spatially uniform resolution, obtained as described above, are denoised by applying a patch-based planar image denoising method. In this chapter, the non-local means (NLM) method [1] is adopted as the patch-based denoising method. The NLM replaces the value of the target pixel with a weighted average of reference pixel values, based on the similarity measure between the block centered on the pixel to be processed and the surrounding reference pixel values. This method assumes uniformity in the spatial resolution of the image. Therefore, this assumption makes its unsuitable for directly applying the NLM method to 360-degree images represented in latitude and longitude formats. In the proposed method, the NLM method computes patch similarity not only between pixels within the same decomposed image, but also across all six decomposed images, as shown in **Figure 7**. Additionally, the NLM method has four parameters: the smoothing parameter, patch size, search

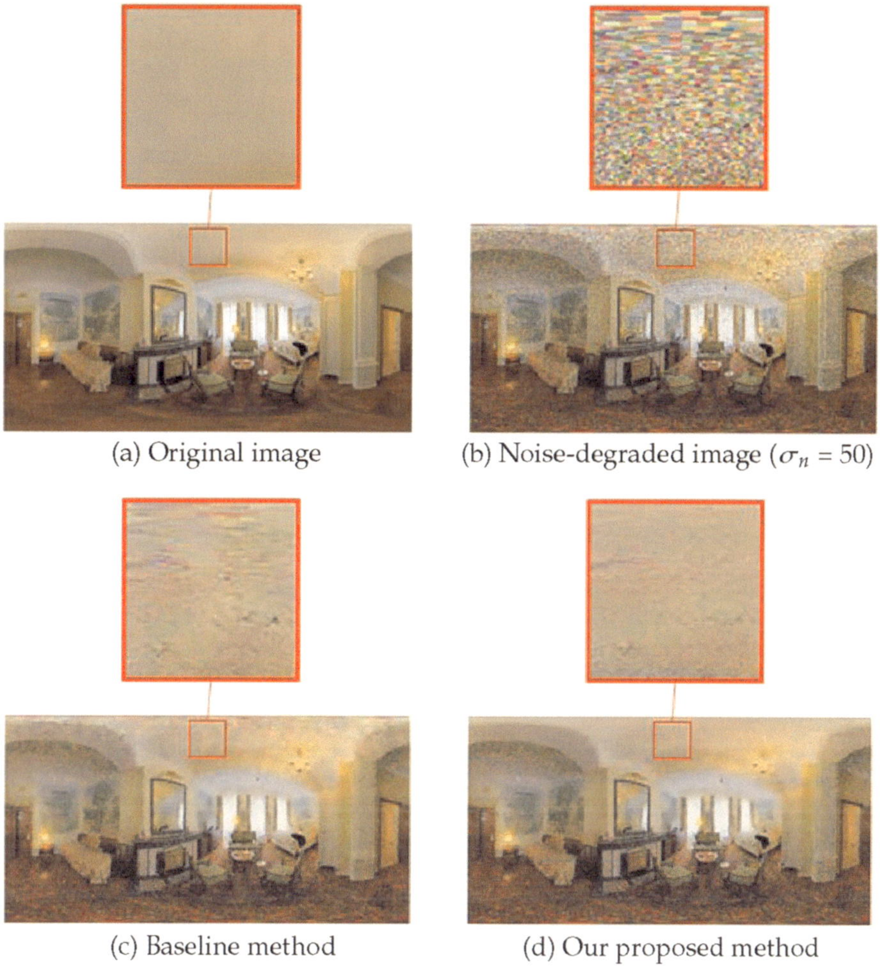

Figure 12.
Visual comparison of 360-degree image denoising on image "5" with noise level of 50. (a) Original and enlarged images. (b) Degraded and enlarged images by Gaussian noise $N(0, 50)$ added from all directions. (c) Denoised and enlarged images with the baseline method. (d) Denoised and enlarged images with the proposed method.

window size, and the standard deviation (SD) of image noise. The optimal values of these parameters are determined using the framework outlined in [12].

To summarize the overall procedures, the 360-degree image is first decomposed into six images with spatially uniform resolution using the method in the previous section. Next, the decomposed images are denoised using the NLM method. Finally, the denoised images are combined into the single 360-degree image. **Figure 8** shows the overall flow of the method.

4. Experimental results

The latitude-longitude images of 360-degree images sampled from the SPSDataset75 dataset [13], which had Gaussian noise added from all directions, were used as input. The method of directly denoising the latitude-longitude images with the NLM filter was used

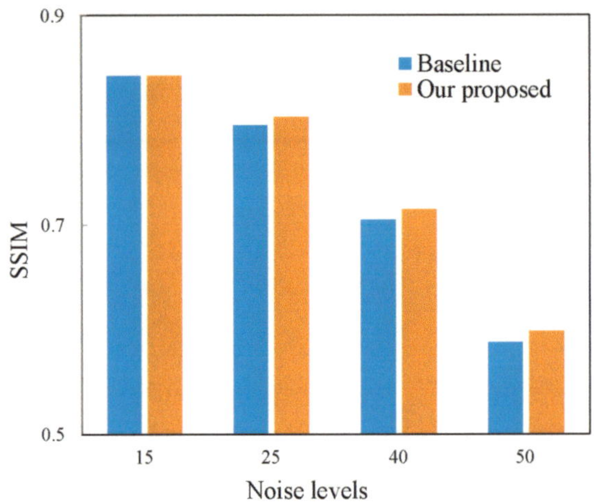

Figure 13.
Structural similarity index measure (SSIM) comparison between baseline and our proposed method.

as the baseline method to compare its performance with our proposed method. The original images from the SPSDataset75 dataset, with an image size of 512 × 1024, and the 360-degree images degraded by Gaussian noise with zero mean and five different standard deviation values ($\sigma_n = 15, 25, 40, 50$) are shown in **Figures 9(a)** and **(b)** to **12(a)** and **(b)**, respectively. The images in (b) from **Figures 9–12**, which are given as input and denoised using the baseline and proposed methods, are shown in images (c) and (d) of **Figures 9–12**, respectively. The enlarged images of the partial regions from **Figures 9–12** show that the proposed method provides higher-quality denoising results than the baseline method for all values of standard deviation.

Figure 13 shows the average SSIM values for the 360-degree image denoising results between the baseline and proposed methods. SSIM measures the similarity between the original and denoised images, with values closer to 1 indicating better performance [14]. Although there is no SSIM performance difference between the two methods, for a noise standard deviation of 15, the difference increases with higher noise levels. The proposed method outperforms the baseline method in SSIM for 360-degree images from the SPSDataset75 dataset.

5. Conclusions

In this chapter, we presented a simple method for extending patch-based denoising algorithms for planar images to 360-degree spherical images. The method is based on the multi-decomposition of 360-degree images, achieved through a novel overset grid method that improves the Yin-Yang grid, providing a more uniform spatial resolution. Experimental results showed both qualitatively and quantitatively that the method effectively denoises 360-degree images. Future work will focus on extending non-patch-based denoising methods to 360-degree images.

Acknowledgements

This research was supported by JSPS KAKENHI Grant Number JP3K11149, Japan.

Author details

Kenji Hara*† and Kohei Inoue†
Department of Media Design, Kyushu University, Fukuoka, Japan

*Address all correspondence to: hara@design.kyushu-u.ac.jp

† These authors contributed equally.

References

[1] Buades A, Coll B, Morel J-M. A non-local algorithm for image denoising. In: Proceedings of 2005 IEEE Computer Society Conference on Computer Vision and Pattern Recognition; 20–25 June. San Diego, CA, USA: IEEE; 2005. pp. 1063-6919

[2] Dabov K, Foi A, Katkovnik V, Egiazarian K. Image denoising by sparse 3D transform-domain collaborative filtering. IEEE Transactions on Image Processing. 2007;**16**(8):2080-2095. DOI: 10.1109/TIP.2007.901238

[3] Mairal J, Bach F, Ponce J, Sapiro G, Zisserman A. Non-local sparse models for image restoration. In: Proceedings of the IEEE International Conference on Computer Vision; 29 September - 2 October; Kyoto. Japan: IEEE; 2009. pp. 2272-2279

[4] Shi H, Traonmilin Y, Aujol J-F. Compressive learning for patch-based image denoising. SIAM Journal on Imaging Sciences. 2022;**15**(3):1184-1212. DOI: 10.1137/21M1459812

[5] Tabassum S, Gowre SC. Optimal image denoising using patch-based convolutional neural network architecture. Multimedia Tools and Applications. 2023;**82**:29805-29821. DOI: 10.1007/s11042-023-15014-8

[6] Zhao W, Riot P, Deledalle C-A, Maître H, Nicolas J-M, Tupin F. Patch-based adaptive temporal filter and residual evaluation. arXiv preprint arXiv: 2402.09561. 2024

[7] Kageyama A, Sato T. Interactive browsing via diversified visual summarization for image search results. Multimedia Systems. 2011;**17**(5):379-391. DOI: 10.1007/S00530-010-0224-7

[8] Purser RJ. The bi-mercator grid as a global framework for numerical weather prediction. In: Proceedings of the 2004 Workshop on the Solution of Partial Differential Equations on the Sphere; 20–23 June 2004. Yokohama, Japan: Frontier Research Center for Global Change; 2004

[9] Hara K, Inoue K, Urahama K. Gradient operators for feature extraction from omnidirectional panoramic images. Pattern Recognition Letters. 2015;**54**(1):89-96. DOI: 10.1016/j.patrec.2014.12.010

[10] Okazaki D, Yu A-S, Hara K. Omnidirectional saliency map generation by Yin-Yang grid method. In: Proceedings of the 2018 2nd International Conference on Video and Image Processing; 29–31 December 2018; Hong Kong. Hong Kong: ACM; 2018. pp. 108-111

[11] Ishihara Y, Inoue K, Hara K. Omni-directional image superpixel segmentation using Yin-Yang grid method. In: Proceedings of the 27th Meeting on Image Recognition and Understanding (in Japanese); 6–9 August 2024; Kumamoto. Japan: IEICE; 2024. p. IS-2-131

[12] Hara K, Inoue K, Urahama K. Full-reference metric adaptive image denoising. In: Proceedings of 2019 IEEE International Conference on Image Processing; 22–25 September 2019. Taipei, Taiwan: IEEE; 2009. pp. 2419-2423

[13] Wan L, Xu X, Zhao Q, Feng W. Spherical superpixels: Benchmark and evaluation. In: Proceedings of the 14th Asian Conference on Computer Vision; 2–6 December; Perth. Australia: Springer; 2018. pp. 703-717

[14] Wang Z, Bovik AC, Sheikh H, Simoncelli EP. Image quality assessment: From error visibility to structural similarity. IEEE Transactions on Image Processing. 2004;**13**(4):600-612. DOI: 10.1109/TIP.2003.819861

Chapter 2

Perspective Chapter: Advancement in CCTV Video Analytics – Leveraging Face Recognition for Enhanced Security Solutions

Fadhil Hidayat, Ulva Elviani and Figo Agil Alunjati

Abstract

In recent years, there has been a significant increase in the integration of CCTV video analytics, particularly in the field of facial recognition technology, aimed at strengthening security measures. This chapter explores the latest advancements in CCTV video analytics and its applications, specifically in the domain of facial recognition. The chapter begins by outlining the fundamental principles underlying CCTV video analytics and the evolution of facial recognition systems, including feature extraction, matching techniques, and in-depth learning methodologies. Additionally, the chapter discusses the practical implementation of CCTV video analytics for facial recognition in various contexts, such as law enforcement, public security, and commercial enterprises. It examines the potential benefits and challenges associated with the deployment of such systems, including privacy concerns and algorithmic biases. Moreover, the chapter highlights emerging trends and future directions in CCTV video analytics for facial recognition, including the integration of multimodal biometrics and the development of real-time surveillance capabilities. This chapter offers a detailed examination of the latest advancements and applications in CCTV video analytics for facial recognition. It provides valuable insights into their potential effects on security infrastructure and explores the broader social implications.

Keywords: face recognition, video analytics, security system, smart system, facial recognition

1. Introduction

Digital integration that continues to develop in the 5.0 era provides significant transformation in various aspects. One transformation in the security system is using CCTV (Closed-Circuit Television) technology. Currently, CCTV has evolved into a smart system that is capable of performing real-time video analysis through the integration of advanced technologies such as machine learning and artificial intelligence. Through a smart system, it allows CCTV to recognize, analyze, and respond to various situations more precisely and intelligently. The concept of a smart system in video analytics basically refers to the use of advanced technology to improve the ability of the system to not only provide output but also provide significant feedback on the problems being solved.

One of the most significant advances in this field is the application of face recognition technology, which is able to provide more effective and efficient security solutions. The increase in the number of crimes and security threats that occur in various sectors, both public and private, requires the right solution to record and provide preventive measures. The ability of face recognition to identify individuals can be one of the preventive measures against case studies that require individual verification, such as attendance, access control, and others.

The combination of CCTV and face recognition by utilizing complex algorithms can speed up the investigation process if a crime occurs. Through this process, the system can recognize individuals who are identified as threats or have a criminal history, thus providing early warning to security officers. The accuracy and speed of this identification process show the great potential of facial recognition technology in improving the quality and effectiveness of security systems, making it a key component in future smart security systems. However, there are still several challenges in implementing this technology, both technically and ethically. Technically, challenges such as lighting conditions, shooting angles, and image quality are still obstacles in achieving maximum accuracy levels. CCTV images and videos require different algorithms compared to broadcast ones due to several key factors:

- *Resolution and quality*: CCTV footage often has lower resolution and quality compared to broadcast videos. This affects the type of algorithms used for tasks like object detection and recognition, as lower-quality images may require more robust preprocessing and noise reduction techniques.

- *Frame rate*: CCTV cameras typically record at lower frame rates than broadcast cameras. This can impact motion detection and tracking algorithms, which need to be optimized for fewer frames per second.

- *Lighting conditions*: CCTV cameras often operate in varying and challenging lighting conditions, such as low-light or high-contrast environments. Algorithms for CCTV need to handle these variations effectively, whereas broadcast videos usually have more controlled lighting.

- *Angle and perspective*: CCTV cameras are often placed in fixed positions with wide-angle lenses, capturing scenes from unusual angles. This requires algorithms that can handle perspective distortion and occlusions, unlike broadcast videos, which are usually shot with optimal angles and perspectives.

- *Purpose and context*: The primary purpose of CCTV is surveillance and security, which involves specific tasks like anomaly detection, behavior analysis, and real-time monitoring. Broadcast videos, on the other hand, focus on delivering high-quality visual content for entertainment or information, requiring different sets of algorithms for editing, color correction, and visual effects.

These differences necessitate specialized algorithms tailored to the unique challenges and requirements of CCTV footage.

Meanwhile, from an ethical perspective, issues of privacy and data misuse are major concerns in the implementation of facial recognition technology. This book explicitly reviews the progress of CCTV video analytics with a focus on facial recognition technology, challenges, and potential for better development in the future in order to create integrated and reliable security solutions.

2. Fundamental principles of CCTV video analytics

German engineer Walter Bruch created closed-circuit television (CCTV) in 1942 for the purpose of monitoring rocket launches after Russian scientist Leon Theremin created it in 1927 for visitor surveillance [1]. Unlike broadcast television, which distributes signals openly, closed-circuit television (CCTV) uses a video camera to deliver television signals to a specific limited channel (a port selected by the owner or camera operator) [1]. Due to its affordability, ease of use, and advantages, CCTV has become an invention that has rapidly expanded throughout society. These days, it is typical to have CCTV installed in homes as well as public areas. This is closely linked to the need for environmental monitoring—which may involve the deployment of CCTV—in order to comply with regulations.

With artificial intelligence starting to have an impact on CCTV, surveillance that was previously limited to human intervention can now be automated and customized to specific video analytics requirements. CCTV videos are sent to object detection systems so they can be analyzed. In general, the video analytics approach using additional visual sources, as shown in **Figure 1**, is comparable to the artificial intelligence (computer vision) process on CCTV cameras. Simply said, there are special characteristics when utilizing CCTV as input.

2.1 Data sensing

This phase focuses on the artificial intelligence system's ability to receive visual stream input from CCTV. The input stream technique depends on a number of variables, including the network protocol and kind of CCTV. Usually, an IP cam is the kind of CCTV that is employed. A local area network (LAN) cable is used by Internet Protocol Cameras, a form of digital video camera, to transmit and receive visual data. IP surveillance on your home Wi-Fi network, CCTV makes use of the same internet infrastructure as other devices [2]. These days, IP cameras are widely employed in many different contexts because of their affordability, convenience of use, and high accessibility, all of which allow for remote monitoring (**Figure 2**). The protocol in use is the second factor.

Numerous delivery protocols are available, including FTP, SMTP, RTSP, HTTP/HTTPS, SIP, and SMTP. Real-Time Streaming Protocol, or RTSP, is a streaming protocol that can be used for low-latency real-time monitoring. However, it has the

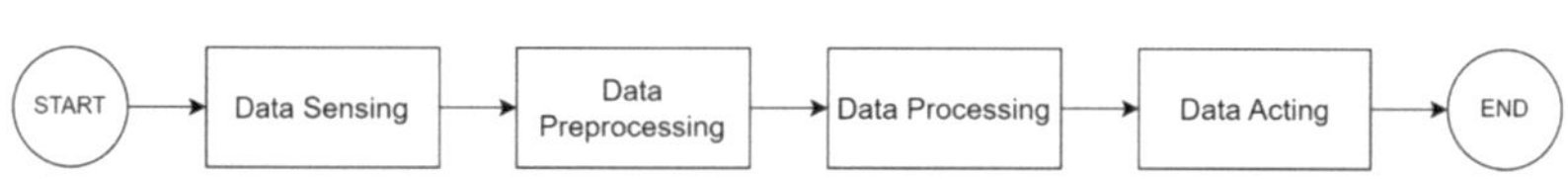

Figure 1.
CCTV video analytics methods.

Figure 2.
Example of installing CCTV indoors.

drawback of continuously delivering data in one direction, which strains the system and may even cause the system software to crash. However, because Hypertext Transfer Protocol (HTTP) employs two-way delivery and does not send streams continually without the system's permission to complete processing the prior stream, it has a better level of reliability. However, there is a considerable latency when using the HTTP protocol.

2.2 Data preprocessing

The main goal of this step is to configure the CCTV input to suit the specific requirements of the system. These adjustments can be tailored to meet specific needs, ensuring the system functions optimally. Examples include whether the system needs a high level of real-time, requires a high video resolution, requires a high FPS rate, etc. Some modifications include color correction, frame enhancement, and resolution changes, all of which contribute to the quality and performance of the video as required. FFMPEG, a versatile tool for handling multimedia data, can be used to efficiently modify these settings. Additionally, FFMPEG can be used to convert the RTSP protocol to HTTP, making the stream less taxing on the system and more stable across different environments.

Preprocessing is important to optimize overall system performance, especially in scenarios where the CCTV setup can be resource-intensive. By converting RTSP to HTTP, the system benefits from reduced latency and better compatibility with different network infrastructures. This approach also simplifies integration with existing web-based monitoring systems, making the entire setup more robust and easy to use. Ultimately, these modifications help ensure that the CCTV system delivers a high-quality and reliable video stream that meets the desired requirements.

2.3 Data processing

This is the primary stage of video analytics, where the necessary features are extracted from the images to produce the desired information. The following are a few of the most popular techniques.

- *Object classification* is a process that involves looking at each frame's visuals and determining whether the object class in that frame and the dataset under study are similar.

- *Object detection* is the process of locating an object in a frame by using classification information.

- *Object tracking* follows each movement of an object's location as it appears in each frame sequentially using detection information.

Depending on the requirements, a variety of algorithms can be applied in data processing. Deep learning algorithms for object recognition and classification range from single-stage detectors like YOLO, RetinaNet, and Single Shot Multi Box Detector to two-stage detectors like the Neural Network Family (CNN, RCNN, etc.). Although they require more processing time, two-stage detectors offer more precision. Single-stage detectors, on the other hand, are appropriate for real-time video analytics and have faster processing times but only moderate accuracy.

2.4 Data acting

At this stage, the main focus is on the system's response to the processed data. This response can take many forms, such as triggering alerts, generating data visualizations, granting access permissions, and more. For example, in a case study involving facial recognition technology, the system can track people's movements, record attendance, or even automatically open gates based on the recognition results. Such actions are critical to automating processes and improving security protocols, making the system efficient and responsive to real-time data.

Other practical examples include detecting suspicious activity, which may prompt the system to send a notification, take a picture, or sound a physical alarm. These reactions are designed to address specific scenarios, ensuring that the system not only identifies potential issues but also responds quickly and appropriately. By integrating these features, the system becomes a powerful tool for monitoring and managing the environment, offering a proactive approach to security and operational efficiency. Each case study highlights the flexibility of the system, showing how it can be tailored to meet a variety of needs and situations effectively.

3. CCTV analytics for facial recognition system

Face recognition is a method used to identify or verify an individual's face from an image or video frame. This technology has emerged as a leading biometric solution due to significant advancements in digital cameras, the internet, and mobile devices, alongside growing security needs. Face recognition offers several benefits over other biometric technologies, including being natural, non-intrusive, and straightforward to implement.

The facial recognition system utilizing CCTV video analytics has several important elements in its development, such as hardware and software components, algorithms and techniques, and how the system is implemented in the real world. This section briefly describes the points regarding the components, algorithms, and case study examples of the application of the face recognition system in the real world.

3.1 Components of facial recognition system

The facial recognition system in the context of CCTV analytics has four interrelated components, such as CCTV cameras and hardware, facial recognition software, facial databases and data management, and network connections and infrastructure. The camera plays an important role in capturing images and videos to be analyzed, where the quality of the images produced by the camera greatly affects the level of accuracy of the facial recognition system. Meanwhile, the success of the software in detecting faces depends heavily on the accuracy of the algorithms and techniques used. The facial database provides a reference for matching detected faces, with good data management being the key to speed and accuracy of matching, as well as privacy protection. Fast network connections and strong IT infrastructure allow the entire system to work in an integrated manner, with a reliable network and scalable storage solutions to handle large volumes of data. Through proper integration of all these components, the facial recognition system in CCTV analytics can operate with high efficiency, providing a sophisticated and reliable security solution.

A system incorporating the face recognition feature generally consists of four modules, as depicted in **Figure 3**, which illustrates the face recognition process: face detection, face alignment, feature extraction, and feature matching.

The initial step in the face recognition process is face detection, wherein the system employs algorithms such as Haar cascades and Histogram of Oriented Gradients (HOG) to identify and locate faces within images or videos. Following detection, the subsequent step is face alignment, which adjusts the orientation of the face to ensure that facial features such as the eyes, nose, and mouth are consistently positioned. The aligned face then undergoes feature extraction, where key characteristics that distinguish one face from another are extracted. The final step is feature matching, which involves comparing the feature vector from the input face image with the feature vectors in the existing database. This process determines whether the identified face corresponds to a face in the database, thereby verifying or identifying the individual.

3.2 Algorithm and technique

There are many algorithms and techniques that can be used in facial recognition systems. There are at least 13 techniques and algorithms that have been summarized in the journal "face recognition for automatic border control: A systematic literature review" [3]. Based on this summary, deep learning and Convolutional Neural Networks (CNN) are the two methods that attract the most interest from researchers to use them (**Figure 4**). This may be due to the use of deep learning, especially Convolutional Neural Networks (CNNs), which have proven to be very effective in recognizing complex visual patterns such as human faces. CNN processes images in increasingly abstract layers, allowing the system to identify even small variations in facial features.

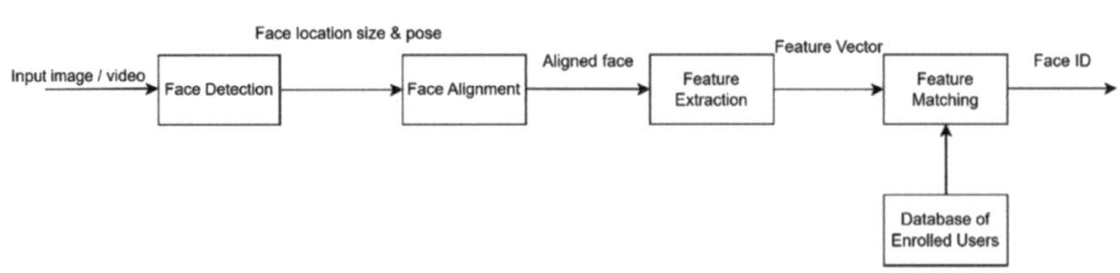

Figure 3.
Face recognition process flow.

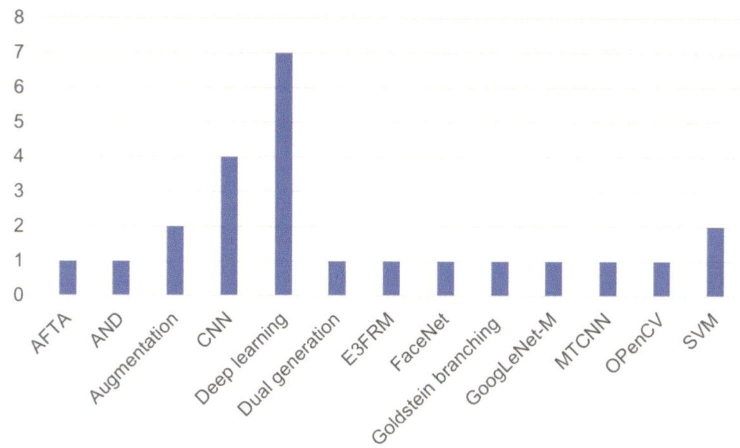

Figure 4.
Face recognition algorithm and technique.

In machine learning, a confusion matrix is commonly employed to determine the success rate of classification based on algorithms or human observations and then compared with accurate measurements. **Figure 5** provides an overview of the confusion matrix for binary classification. The confusion matrix is divided into two components: the actual class and the predicted class. The actual class represents the true condition of the data being analyzed, while the predicted class reflects the outcomes generated by the prediction process.

Classification performance metrics are essential for evaluating the efficacy of test results. **Table 1** illustrates several calculations within the Classification Performance Matrix, namely sensitivity, specificity, accuracy, precision, and the F1 score. Sensitivity, also referred to as the True Positive rate, quantifies the proportion of actual positives correctly identified in the testing phase. Specificity, or the True Negative rate, measures the proportion of actual negatives accurately detected. Accuracy encompasses both the True Positive and True Negative rates in relation to the total number of tests conducted. Precision, defined as the positive predictive

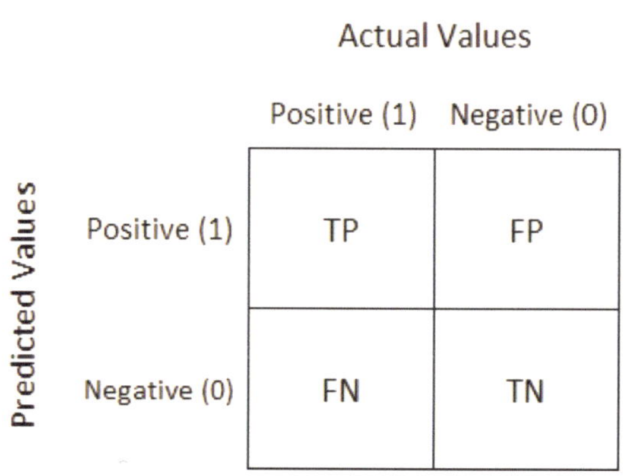

Figure 5.
Confusion matrix for binary classification, TP: True Positive, FP: False Positive, FN: False Negative, TN: True Negative.

Symbol	Matrix	Formula
SNS	Sensitivity	TP/(TP + FN)
SPC	Specificity	TN/(TN + FP)
ACC	Accuracy	(TP + TN)/total data
PRC	Precision	TP/(TP + FP)
F1	F1 Score	2*SNS*PRC/(SNS + PRC)

Table 1.
Classification performance matrix.

value, indicates the proportion of positive results that are accurately identified. The F1 score represents a weighted average of precision and sensitivity, providing a balanced assessment of the classification performance.

3.3 Implementation of CCTV analytics for facial recognition system

Basically, the implementation of CCTV analytics for facial recognition systems requires careful planning and selection of the right technology to achieve optimal results. One example of the application of face recognition is in automatic border control machines. The European Union started implementing this technology in the early 2010s, which of course is one of the first major initiatives in the history of implementing facial recognition systems at airports. The reason for this implementation is to overcome various problems that occur at border control, which often passes inspections on subjects registered on the watchlist or blacklist. Document matching does not guarantee that the documents provided are genuine. Data on the blocklist also does not have adequate and up-to-date information about a person's identification; moreover, a person's face can change with age, making identity and recognition difficult [4]. The use of cosmetic products can also cover certain facial features. Plastic surgery can also change a person's appearance. Wearing a headscarf for religious reasons can make identifying a person based on their face more difficult in some countries because it covers certain facial features. This weakness causes inaccurate facial identification and recognition in border control. The application of facial recognition systems in automatic border control is not only a verification step but also a preventive measure against various threats such as presentation attacks [5], morphing attacks [6, 7], adversarial attacks [8], live spoofing [9], inconsistent matching rates due to certain factors or network availability issues [10].

4. Challenges and future direction

4.1 Facial recognition methodology

Facial recognition methodologies vary widely, shaped by the range of available data sets and specific needs. Security systems, for example, often prioritize accuracy and reliability to ensure proper identification, while real-time-focused applications lean toward speed and a seamless user experience. Depending on these goals, techniques such as deep learning or more traditional feature-based methods are used. The challenge is finding the right balance—choosing a methodology that not only meets accuracy standards but also maintains efficiency and scalability across scenarios. This balancing act underscores the need for ongoing research and innovation to keep pace with changing demands and emerging threats.

As technology evolves, so do the requirements for facial recognition systems, driving a continuous cycle of improvement and adaptation. New data sets, often more complex and diverse, require refined algorithms that can handle greater variability in facial features and environmental conditions. This evolution demands a nuanced understanding of the strengths and limitations of current methodologies. Researchers and developers must remain agile, ready to integrate new techniques that offer better performance in real-world conditions. Ultimately, the goal is to create a system that is not only effective in today's environment but also resilient and adaptable to future challenges, ensuring long-term reliability and security.

4.2 Privacy

Privacy concerns in facial recognition technology are particularly important because facial data is the primary biometric identifier, making it highly sensitive and vulnerable to misuse. Issues such as unauthorized surveillance, data breaches, and potential identity theft have become major topics of ethical and legal discussion. To address these concerns, regulatory frameworks such as the General Data Protection Regulation (GDPR) have been established, which set strict guidelines on how facial data is collected, stored, and used. These regulations aim to protect individuals' privacy rights while allowing the technology to continue to evolve. However, finding the right balance between leveraging the security benefits of facial recognition and ensuring strong privacy protections remains a complex challenge.

In response to these concerns, there is a growing focus on the development of privacy-preserving algorithms and secure biometric systems. These emerging technologies aim to mitigate privacy risks by implementing measures such as data anonymization, encryption, and decentralized storage. The goal is to enable the continued use of facial recognition in a variety of applications without compromising individual privacy. As the technology evolves, strategies for protecting personal data must also evolve, ensuring that advances in facial recognition do not come at the expense of ethical considerations. This ongoing effort requires collaboration between technologists, legal experts, and policymakers to create systems that are effective and respect privacy rights.

4.3 Large-scale face recognition

The scalability of facial recognition systems to manage large datasets and handle high query volumes involves overcoming significant technical hurdles. Efficient indexing and retrieval algorithms play a critical role in ensuring that the identification process remains fast, even when dealing with millions of facial records. To cope with the enormous processing demands, distributed computing architectures are often used, spreading the workload across multiple systems to optimize performance. However, maintaining accuracy while scaling, managing large storage requirements, and ensuring real-time response are ongoing challenges that require careful balancing.

Advances in cloud computing and parallel processing technologies are pushing the boundaries of what is possible in large-scale facial recognition systems. These innovations are focused on improving the efficiency and scalability of these systems, making them more adaptable to a variety of operational settings. As these systems evolve, so does the complexity associated with them, requiring continuous refinement of algorithms and infrastructure to meet the demands of different applications. The future of large-scale facial recognition lies in the seamless integration of all these technologies, ensuring that systems remain robust and precise as they expand to accommodate larger and more diverse datasets.

4.4 Face variability

Recognizing faces across different poses (e.g., frontal, profile) and varying facial attributes (such as glasses, masks, or hijabs) is a complex challenge for face recognition systems. Techniques such as 3D modeling and multi-view learning are utilized to enhance recognition accuracy under pose variations. Algorithms must also handle changes in appearance due to accessories or clothing, ensuring robustness against occlusions and variations in facial expressions. Addressing these challenges involves developing adaptive algorithms capable of effectively recognizing individuals despite diverse facial characteristics and conditions.

Research on the impact of pose variation and facial diversity has been conducted by [11]. This study performed camera angle configurations to produce facial datasets that produce the most optimal facial recognition accuracy results. This study produced the Tilt-angle Face Dataset (TFD), containing 11,124 facial images from 927 subjects covering various facial tilt angles. The results of testing in this study are that the accuracy of facial recognition using a combination of the TFD dataset with WebFace significantly outperforms the results using only the WebFace dataset for training. An accuracy of 91.65% success was obtained using a training model with a combination of TFD and WebFace.

Similar research investigates how different conditions in capturing facial datasets affect the performance of facial recognition systems, focusing on factors like pose changes and external factors such as varying lighting conditions and image resolution. The dataset was collected under varying conditions, with and without specific poses (e.g., consistently facing the camera). **Figure 6** shows the capture of a facial dataset with poses. The study also employed multiple camera angles to capture different sides of the face, aiming to thoroughly explore facial features.

The evaluation of the datasets was carried out using the ArcFace and FaceNet models. Testing results indicated that the training dataset with lighting at 690 lux, a resolution of 1280 × 720 pixels, and pose variations produced the most reliable performance for ArcFace in attendance systems compared to other datasets. Moreover, it was concluded that dim lighting conditions in the training dataset could negatively impact ArcFace's facial recognition results. Pose variations helped improve

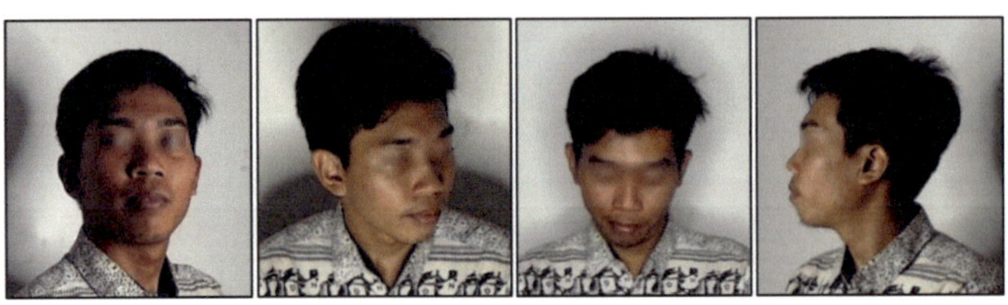

Figure 6.
Pose variations in dataset acquisition.

the performance of the ArcFace model by reducing false recognitions, and variations in image resolution did not significantly affect the model's performance. **Figure 7** illustrates how facial attributes like hijabs, masks, glasses, or head coverings present challenges in improving face recognition accuracy, a challenge that will continue to evolve with advancements in fashion and technology.

4.5 Environment variability (lighting, camera resolution)

Face recognition performance can be significantly influenced by environmental factors such as lighting conditions and camera resolutions. Algorithms need to adjust to variations in lighting that can obscure or alter facial features, while high-resolution cameras capture detailed information at the cost of increased computational demands and storage requirements. Adaptive algorithms are crucial to maintaining accuracy across different environmental conditions by adjusting recognition parameters based on environmental cues. Ongoing research aims to improve algorithms' robustness to environmental variability, ensuring reliable performance in diverse settings. Face recognition performance can be significantly influenced by environmental factors such as lighting conditions and camera resolutions. Algorithms need to adjust to variations in lighting that can obscure or alter facial features, while high-resolution cameras capture detailed information at the cost of increased computational demands and storage requirements. Adaptive algorithms are crucial to maintaining accuracy across different environmental conditions by adjusting recognition parameters based on environmental cues. Ongoing research aims to improve algorithms' robustness to environmental variability, ensuring reliable performance in diverse settings.

The environment greatly influences the development of a face recognition system, one of which is for access control, such as camera placement and lighting. Qian Zhai's 2022 study on face video detection for access control systems aimed to enhance information management, face feature monitoring, and recognition accuracy in environments with a random flow of people [11]. The research optimized system design by integrating image processing, computer vision, and machine learning to better detect and identify human characteristics. These processes are carried out to obtain optimal performance with an unusual camera position. Key advancements included block matching for video features, edge contour analysis, and biological directional recognition, along with a face anti-deception algorithm for improved authentication.

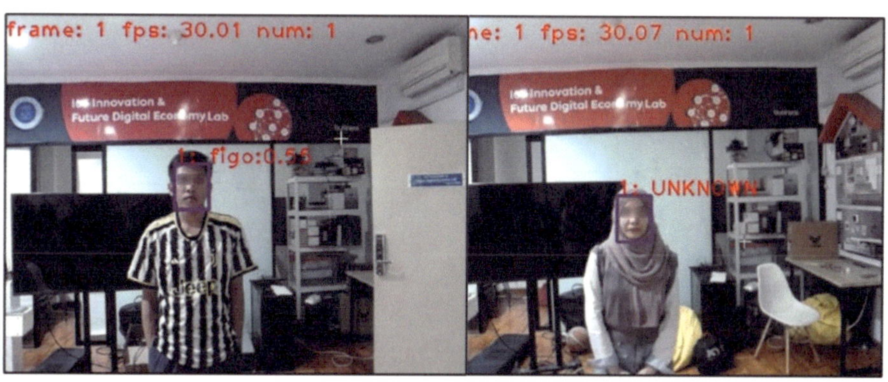

Figure 7.
The effect of face variability on accuracy.

The proposed system significantly outperformed traditional methods, achieving an accuracy of 0.932 compared to 0.684 in 100 iterations. These improvements highlight the system's potential for more reliable and secure access control in complex environments.

Similar research focuses on the impact of environmental factors on face recognition accuracy in an access control system for a mining construction site. The uncontrolled nature of the mining construction environment poses unique challenges to face recognition accuracy. The study outlines several environmental factors, such as extremely bright lighting that casts shadows on faces, the high positioning of cameras due to manufacturing frames, and weather changes like rain that cause blurring.

The dynamic outdoor mining environment significantly affects test results, yielding an accuracy of 60%, precision of 96%, recall of 58%, and an F-score of 72% in a representative simulation space. **Figure 8** provides examples of test outcomes across various scenarios, including worker arrivals, mining activities, and worker discussions. The system developed in this study aims to track worker attendance and monitor activities at a mining site.

4.6 Face occlusion

Facial occlusion, where parts of the face are obstructed or covered, significantly complicates the task of accurate facial recognition. Scenarios involving individuals wearing masks or other partial face coverings are becoming increasingly common, requiring systems to adapt to these challenges. Techniques such as partial matching, which uses only visible facial features, and holistic recognition, which attempts to infer identity from the entire context, are critical in overcoming the limitations caused by occlusion. The goal is to ensure that even when facial data is incomplete, the system can still make reliable identifications.

Figure 8.
Face detection accuracy results in mining simulation.

To address these challenges, researchers have focused on developing algorithms that can effectively process missing or unobserved facial features. Advances in deep learning, particularly in the areas of image inpainting and feature reconstruction, show promise in reconstructing occluded facial regions, thereby improving overall recognition accuracy. These innovations are critical in making facial recognition systems more robust and reliable in real-world applications where facial occlusion is common. As these technologies mature, they have the potential to significantly improve the effectiveness of facial recognition in diverse and challenging environments.

4.7 Multimodal biometric system

Multimodal biometric systems integrate multiple biometric identifiers, such as face, fingerprints, and voice, to enhance overall accuracy and reliability. By combining modalities, these systems mitigate individual limitations and improve robustness against spoofing attacks. Challenges include synchronizing data from disparate sources and ensuring interoperability among different biometric technologies. Machine learning techniques play a crucial role in integrating and fusing multimodal inputs to make informed decisions. Advancements in these systems aim to create seamless and secure authentication solutions adaptable to various operational environments, enhancing user convenience and security.

The use of multimodal biometric systems is essential due to various biometric spoofing techniques, such as face morphing. A literature review conducted by researchers, as detailed [12], focused on the issue of face morphing, particularly in the context of immigration. Face morphing involves altering facial images to create fraudulent identities, using tools or software that are widely available, both free and paid. If undetected, it could lead to the issuance of passports to unauthorized individuals, posing significant security risks. The purpose of this analysis is to understand face morphing and blacklist attempts in passport applications, providing a safeguard for immigration officers.

Face morph detection methods can be categorized based on the scenarios they address. For single-image-based detection, these methods include texture-based, quality-based, noise-based, deep learning-based, and hybrid approaches. Texture-based analysis examines the texture features of photos, while quality-based methods assess image quality by measuring distortion or degradation to detect morphs. Noise-based methods analyze pixel anomalies that may occur during the morphing process. Deep learning-based approaches utilize deep learning techniques for image classification. Hybrid methods combine multiple features or classifications to detect face morphs. The findings suggest that the developed face morph detection systems show promise for enhancing passport issuance security, with challenges that may be addressed in future research.

5. Closing remarks

In summary, the advancements in CCTV video analytics, especially in facial recognition technology, signify a pivotal shift in security infrastructure across various sectors. By delving into the core principles and evolution of these technologies, this chapter has provided a thorough understanding of their underlying mechanisms, from feature extraction to sophisticated deep learning methodologies. The practical applications in law enforcement, public security, and commercial enterprises underscore the transformative potential of CCTV video analytics in enhancing safety and operational efficiency. However, the deployment of these technologies is not without its challenges. Privacy concerns and algorithmic biases present significant hurdles that must be addressed to ensure ethical and equitable use. As we navigate these complexities, it is crucial to develop robust frameworks that balance security needs with individual rights. Looking ahead, the integration of multimodal biometrics and the advancement of real-time surveillance capabilities promise to further revolutionize the field. These emerging trends point toward a future where CCTV video analytics and facial recognition systems are not only more accurate and efficient but also more adaptable to diverse and dynamic environments. Ultimately, the insights provided in this chapter highlight the dual nature of these technologies: their potential to significantly enhance security and their implications for privacy and social dynamics. As stakeholders in this evolving landscape, it is our responsibility to foster innovations that uphold the principles of fairness and respect for all individuals. The future of CCTV video analytics and facial recognition will be shaped by our collective efforts to harness these tools responsibly and ethically.

Acknowledgements

In the compilation of this chapter, we were ably assisted by researchers from the Smart City Community and Innovation Center Bandung Institute od. We extend our profound gratitude to all the researchers for their invaluable contributions.

Conflict of interest

The authors declare that there are no conflicts of interest regarding the publication of this work. Any opinions expressed herein are the author(s)' own and are not influenced by any affiliations or financial relationships with organizations, entities, or individuals that could be perceived as a conflict of interest. Furthermore, the research and content presented in this book have been conducted independently, without any financial support or influence from external sources that may have a vested interest in the outcomes. All sources of funding for the research, if any, have been acknowledged in the appropriate sections of this book. The authors strive to maintain transparency and integrity in their work, ensuring that the information presented is based on objective analysis and reliable data. Any potential conflicts of interest identified during the course of writing and research have been disclosed and addressed to uphold the credibility and trustworthiness of this publication.

Author details

Fadhil Hidayat*, Ulva Elviani and Figo Agil Alunjati
Smart City Community and Innovation Center, Bandung Institute of Technology, Bandung, Indonesia

*Address all correspondence to: fadhil_hidayat@itb.ac.id

References

[1] Martin R. Closed-Circuit Television. Chicago, IL: Encyclopedia Britannica. Available from: https://www.britannica.com/technology/closed-circuit-television; [Accessed: August 14, 2024]

[2] OHEAP. Available from: https://www.oheap.co.uk/insights/what-is-ip-camera-cctv/

[3] Hidayat F, Elviani U, Situmorang GBG, Ramadhan MZ, Alunjati FA, Sucipto RF. Face recognition for automatic border control: A systematic literature review. IEEE Access. 2024;**12**:37288-37309. DOI: 10.1109/ACCESS.2024.3373264

[4] Zhao J, Yan S, Feng J. Towards age-invariant face recognition. IEEE Transactions on Pattern Analysis and Machine Intelligence. 2022;**44**(1):474-487. DOI: 10.1109/TPAMI.2020.3011426

[5] Jacques Junior JCS et al. First impressions: A survey on vision-based apparent personality trait analysis. IEEE Transactions on Affective Computing. 2022;**13**(1):75-95. DOI: 10.1109/TAFFC.2019.2930058

[6] Ortega-Delcampo D, Conde C, Palacios-Alonso D, Cabello E. Border control morphing attack detection with a convolutional neural network de-morphing approach. IEEE Access. 2020;**8**:92301-92313. DOI: 10.1109/ACCESS.2020.2994112

[7] Kraetzer C, Makrushin A, Dittmann J, Hildebrandt M. Potential advantages and limitations of using information fusion in media forensics—A discussion on the example of detecting face morphing attacks. EURASIP Journal on Information Security. 2021;**2021**(1):1-25. DOI: 10.1186/s13635-021-00123-4

[8] Hwang RH, Lin JY, Hsieh SY, Lin HY, Lin CL. Adversarial patch attacks on deep-learning-based face recognition systems using generative adversarial networks. Sensors. 2023;**23**(2):1-29. DOI: 10.3390/s23020853

[9] Huszár VD, Adhikarla VK. Live spoofing detection for automatic human activity recognition applications. Sensors. 2021;**21**(21):1-20. DOI: 10.3390/s21217339

[10] Khan N, Efthymiou M. The use of biometric technology at airports: The case of customs and border protection (CBP). International Journal of Information Management Data Insights. 2021;**1**(2):100049. DOI: 10.1016/j.jjimei.2021.100049

[11] Wang N, Wang Z, He Z, Huang B, Zhou L, Han Z. A tilt-angle face dataset and its validation. In: 2021 IEEE International Conference on Image Processing (ICIP). Anchorage, AK, USA; 2021. pp. 894-898. DOI: 10.1109/ICIP42928.2021.9506052

[12] Sucipto RF, Hidayat F. Face morph detection: A systematic review. In: 2022 International Conference on ICT for Smart Society (ICISS). Bandung, Indonesia; 2022. pp. 1-6. DOI: 10.1109/ICISS55894.2022.9915233

Chapter 3

Analysis of Similarity Structures for Star Identification in Blurred Images

Aline Nunes de Souza, Rodrigo Marques de Figueiredo, Vitor Camargo Nardelli and Jean Schmith

Abstract

Astronomical observations often suffer from environmental variations that cause blurring in telescopes, compromising the quality and sharpness of the captured images. To mitigate these effects, this work proposes a methodology for detecting stars in blurred images by developing a mathematical model based on probabilistic approaches, exploring the relationship between the light dispersion of stars and a Gaussian distribution. The developed model incorporates two approaches to define the similarity structure: correlation and orthogonality similarity. The results obtained did not show false positives when using orthogonality similarity, achieving a precision of 100%, while the correlation method reached a precision of 95%, highlighting the potential of orthogonality similarity as an effective alternative to identify stellar patterns under blurred conditions. For a dataset with homogeneous star luminosity, an accuracy of 100% was achieved.

Keywords: astronomy, star detection algorithm, blurred images, computer vision, similarity metrics

1. Introduction

Astronomical observation, particularly accurate identification of stars, plays a crucial role in predicting natural phenomena such as solar activity and its effects on Earth. In addition, it facilitates the monitoring of climate change, the calibration of scientific instruments, navigation, and orientation in space missions.

Blur, as described by Vijayarani et al. [1], refers to the loss of sharpness in an image, which reduces clarity and detail, resulting in a softened and less defined appearance. In satellite imagery, this effect often comes from motion, thermal-induced lens defocus, or external disturbances, significantly degrading image quality and complicating analysis [2]. For star identification, such noise and blur obscure critical features, posing challenges to accurate detection and further complicating the analysis of celestial objects in observational images affected by environmental conditions or equipment limitations. Existing approaches in the literature address this issue

Article	Methodology	Metrics
Zhu et al. [3]	Ordered Set of Points (OSP), kNN algorithm, hashing table, Fast Search Algorithm	ACC: higher than 98% with centroid error of 3 pixels
Zhao et al. [4]	Karhunen-Loève Transformation (K-L), Star Walk Formation	ACC: 99.1%
Liang et al. [5]	Image Normalization, Zernike Moments	ACC: 99.27%
Zhou et al. [6]	Multilayer voting algorithm combining triangle-based voting and Singular Value Decomposition (SVD)	Not reported
Liu et al. [7]	One-dimensional Gaussian Morphology	Not reported

Table 1.
Summary of star identification algorithms.

using diverse methodologies. **Table 1** provides an overview of the star identification algorithms, highlighting their methodologies and performance metrics. Notable methods include the Karhunen-Loève Transformation and Zernike Moments, which achieve high accuracy rates (over 99%), while others, such as One-dimensional Gaussian Morphology, do not report quantitative metrics. For example, Zhu et al. [3] use ordered point patterns and hash-based searches to improve star identification speed and reliability under centroid errors. Zhao et al. [4] utilize the Karhunen-Loève transform for noise reduction, while Liang et al. [5] incorporate Zernike moments to extract rotation-invariant features. Zhou et al. [6] propose a multilayer voting algorithm combining triangle voting with Singular Value Decomposition (SVD) to improve noise robustness, and Liu et al. [7] applied one-dimensional Gaussian morphology for segmenting stellar objects in complex backgrounds. Despite their efficacy, these methods often require high computational resources or homogeneous datasets with low noise levels.

This work aims to address these limitations by proposing a simpler, computationally efficient algorithm for star identification in noisy images. By estimating the correlation or similarity between stars' luminous dispersion and Gaussian distribution, the methodology eliminates the need for stellar catalogs, enabling integration into devices with limited storage capacity. Key metrics, including true positive rate (TPR), false positive rate (FPR), precision, accuracy (ACC), and recall, were used to evaluate the performance of the proposed algorithm. The proposed approach prioritizes mathematical simplicity and practicality with the aim of achieving robust star identification while addressing challenges such as noise, blur, and computational constraints.

2. Methodology

The methodology employed in this work is summarized through the block diagram in **Figure 1**, which organizes the fundamental processing layers of the star identification model, without detailing the two similarity methods, discussed later in the text. The process begins by loading an image from the dataset, converting it to grayscale, and normalizing the pixel values to the range $[0, 1]$. Next, an intensity filter and binary mask are applied to suppress noise and emphasize bright regions. These preprocessed

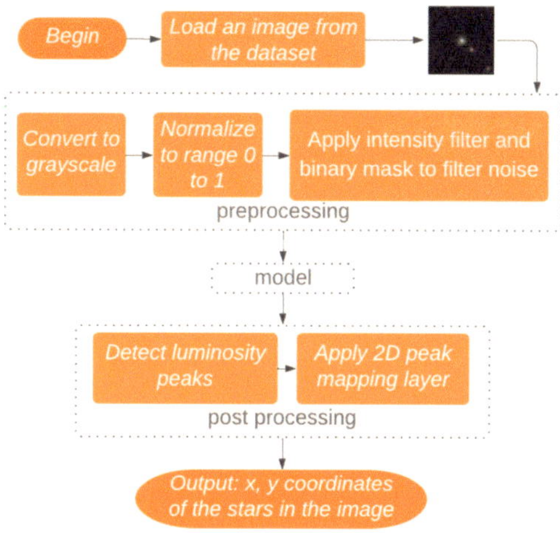

Figure 1.
Block diagram of the proposed method.

images are fed into the model, which detects luminosity peaks and applies a mapping layer to identify potential star candidates. The output consists of the (x, y) coordinates of the detected stars, representing their positions in the image.

2.1 Dataset

For the purpose of validating the star identification model, two datasets containing a variety of astronomical images were selected for validation purposes. The first dataset originates from the Sloan Digital Sky Survey (SDSS) [8], while the second is provided by AGRAWAL [9], which contains images captured by the Devasthal Fast Optical Telescope (DFOT).

2.2 Preprocessing

After selecting the image for detection, preprocessing is performed. First, the RGB images are converted to grayscale. Then, the values are normalized to facilitate processing. Finally, an intensity filter is applied to highlight the brightest regions while removing noise. This filter sets pixel values below 20% of the maximum intensity to zero, effectively reducing low-luminance noise. It also enhances regions of interest by preserving only pixels with significant brightness, ensuring that the model focuses on the most relevant features.

2.3 Model

The central hypothesis of this work is based on previous astronomical observations, which define the luminosity of stars captured by a telescope camera as similar to Gaussian distributions, a phenomenon discussed by Wang et al. [10]. According to Wang et al. [10], the energy distribution of a real star image statistically aligns with a Gaussian distribution in most cases. Accordingly, an image of a star captured by a telescope exhibits a radial decay in luminosity from the center of the captured light.

Figure 2 illustrates this phenomenon with a three-dimensional representation of the grayscale intensity generated from a star image in the SDSS database [8]. The left

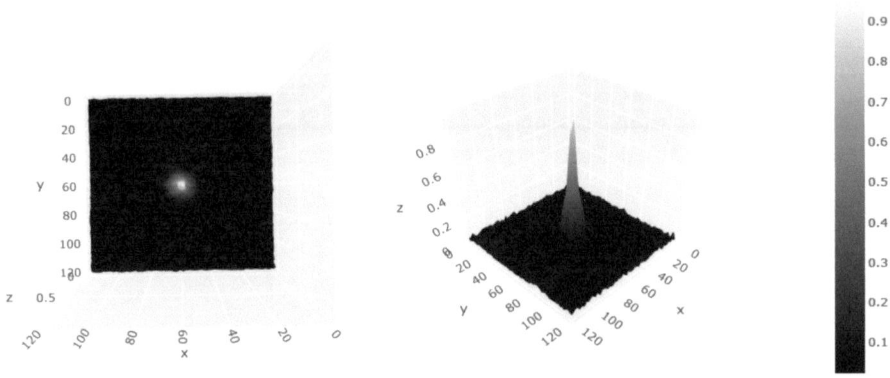

Figure 2.
Light intensity surface of an image with a star.

image displays the original star image in 2D grayscale, highlighting the star's central region with higher intensity. The right image provides a 3D perspective of the same image, visualizing the intensity distribution as a surface plot. In this representation, the peak at the center corresponds to the brightest region of the star, while the gradual decay toward the edges reflects the Gaussian-like distribution. The scales in the figure provide additional context. In the 2D image, the x and y axes represent the spatial dimensions of the image in pixels, while the grayscale color bar on the right indicates the pixel intensity values, ranging from 0 (black) to 1 (white). In the 3D plot, the z-axis represents the normalized intensity values, emphasizing the Gaussian-like peak at the center.

The methodology employed in this work for star detection adopts probabilistic inference approaches, combining correlation and similarity maps between the observed image and two-dimensional kernels. These kernels serve to generate a new image that highlights the points with the highest similarity in terms of their properties, thereby acting as pattern detectors.

2.3.1 Gaussian Kernel

The kernel used in this work follows a Gaussian distribution, consistent with the approach discussed by Wang et al. [10]. It is parameterized by two factors: the kernel size (K_s) and the spread index, represented by the standard deviation (σ), as defined by the discrete Eq. (1).

$$G[i,j|K_s,\sigma] = \frac{1}{2\pi\sigma^2} e^{\left(-\frac{\left(i-\frac{K_s-1}{2}\right)^2+\left(j-\frac{K_s-1}{2}\right)^2}{2\sigma^2}\right)} \tag{1}$$

The first step in developing the model was the application of a Gaussian kernel to the processed image, aiming to smooth the variation in luminosity and enhance the bright patterns that characterize stars. This step involves creating a Gaussian filter which was applied to a sliding window over the image, thereby smoothing and reducing noise. The kernel size and the sigma parameter, which controls the dispersion of the Gaussian function, were adjusted according to the image resolution and the visual characteristics of the stars. The kernel size varies from 3 to the size of the image in increments of 2 (i.e., 3, 5, 7), ensuring that the kernel size is always odd. The value of σ ranges from 1 to $2K_s + 3$, increasing as the kernel size K_s increases. This controls the dispersion of the Gaussian filter, making the image smoothing more pronounced as σ increases. For example, in a 120x120 image, the kernel size ranges from 3 to 199, while σ ranges from 1 to 240. For feature extraction, two methods were tested: orthogonal similarity and correlation.

2.3.2 Orthogonal similarity

The first feature extraction technique, illustrated in the block diagram of **Figure 3**, is based on orthogonality similarity. This method processes an input image using a Gaussian kernel defined by specific parameters K_s and σ. The region of the image that matches the kernel size is extracted and the empty areas are filtered to avoid irrelevant calculations. The orthogonality similarity is then determined between the kernel and the extracted region, followed by applying a binary mask to highlight regions with high similarity. Finally, the similarity maps from multiple kernels are aggregated to produce the final result.

The orthogonality similarity is formally defined by Eq. (2), which evaluates the alignment between two matrices A and B, in the context of this work, the kernel and a region of the image, by normalizing their inner product with their respective magnitudes. The similarity ranges from −1 to 1, where values closer to 1 indicate a higher alignment:

$$OS = \frac{\sum_{i=1}^{M}\sum_{j=1}^{N} A[i,j]B[i,j]}{\sqrt{\sum_{i=1}^{M}\sum_{j=1}^{N} A[i,j]^2}\sqrt{\sum_{i=1}^{M}\sum_{j=1}^{N} B[i,j]^2}} \tag{2}$$

2.3.3 Correlation

The second feature extraction technique, as illustrated in the block diagram in **Figure 4**, is based on the Correlation Index. This method utilizes a Gaussian kernel, defined by its size K_s and standard deviation σ, to calculate a correlation map. The correlation map, mathematically defined in Eq. (3), identifies regions in the image whose intensity distributions closely resemble the shape, size, and dispersion of the kernel.

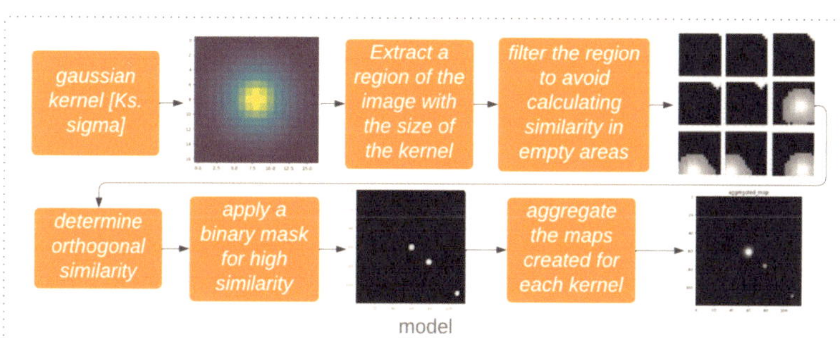

Figure 3.
Block diagram of the model based on orthogonal similarity.

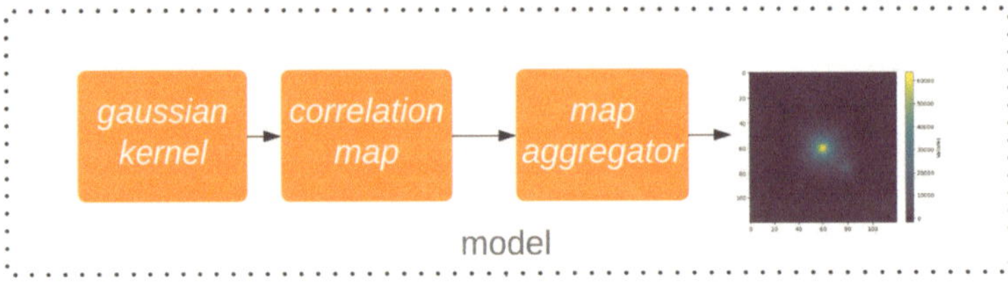

Figure 4.
Block diagram of the model-based correlation.

$$C[x,y|K_s,\sigma] = \sum_{i=0}^{M-1}\sum_{j=0}^{N-1} I[x-i,y-j]G[i,j|K_s,\sigma] \tag{3}$$

The process begins by applying the Gaussian kernel to the input image, where it is systematically slid across all positions. For each position, the weighted sum of overlapping pixel intensities is computed, producing the correlation map. This map highlights areas of the image that share strong similarities with the kernel, thus identifying potential stellar objects.

To account for stars of varying sizes, multiple Gaussian kernels with different scales are applied. This ensures that regions containing larger or smaller stars can still achieve high correlation values. The use of multiple kernels improves the robustness of the detection process, as each kernel is expected to match regions corresponding to stars of similar dimensions.

After generating individual correlation maps for each kernel, a map aggregator combines these results. By computing the arithmetic mean of the correlation maps, the aggregator produces a refined final map that accentuates regions with the highest likelihood of containing stars. This step not only consolidates information, but also reduces noise and artifacts from individual maps.

In summary, this correlation-based approach applies Gaussian kernels to approximate the intensity profiles of stars, enabling the identification of regions that align with the kernel's characteristics. By using multiple kernels and aggregating the results, the method accommodates variations in star sizes and highlights areas with the highest probability of containing stars.

2.4 Post-processing

The first step in the post-processing stage is the detection of brightness peaks. This technique was applied to identify regions of maximum intensity along the horizontal (x) and vertical (y) axes. The image was first converted into a grayscale intensity matrix, where each pixel's brightness was represented numerically. These intensity values were then mapped to their respective x- and y-coordinates, resulting in a two-dimensional distribution of brightness.

Maximum intensity values were extracted along the x (rows) and y (columns) axes, and brightness peaks were identified based on their prominence and height. To ensure the relevance of the detected peaks, thresholds were established: a minimum prominence of 15% of the maximum intensity and a height of at least 55% of the maximum intensity. This method ensured that only the most significant brightness regions were considered.

The detected peaks were visualized in plots, as shown in the example of **Figure 5**, revealing three primary sources of brightness aligned along both axes. Among these,

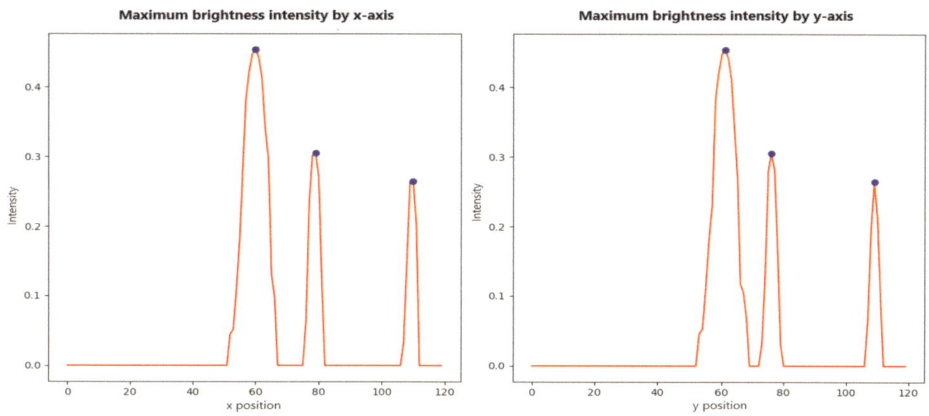

Figure 5.
Brightness intensity graph of the image with stars.

one peak exhibited a notably higher intensity, indicating a dominant light source, likely a brighter or larger star compared to the others.

The peaks were then localized in the original image. After the brightness peaks were detected, their corresponding locations in the original image were identified. The x- and y-coordinates of the peaks were cross-referenced to pinpoint areas where high-intensity values coincided. Only points exceeding 55% of the maximum intensity were considered valid. These points were then marked directly on the original image with small red circles, highlighting potential stars or other high-intensity objects. The final result, displayed in the example of **Figure 6**, illustrates the marked points of interest.

2.5 Validation

After the luminosity peaks were detected, the results were visualized for qualitative analysis. The detected stars were marked on the original astronomical image, allowing a visual inspection of the accuracy of the developed model. Additionally, graphs of the brightness intensity curves in the x and y directions were generated, with the detected peaks highlighted to facilitate the analysis.

To evaluate the accuracy of the detected point coordinates, two datasets were utilized: one composed of non-homogeneous and the other of homogeneous data.

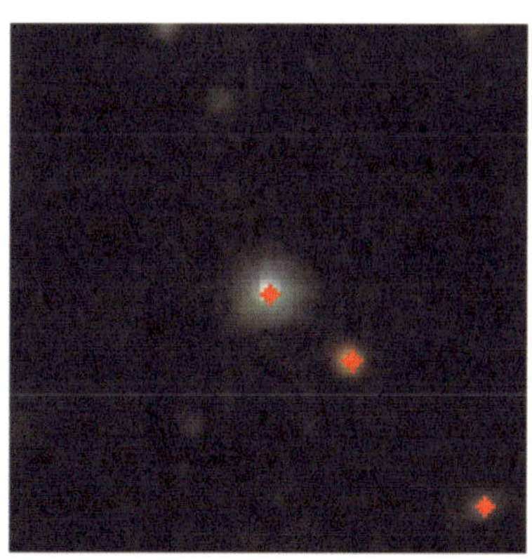

Figure 6.
Original image with candidate star points indicated in red.

The non-homogeneous dataset includes images without discrimination regarding the shape, size, or varying luminosity levels of the stars. This dataset comprised 52 labeled images, eight of which were sourced from the dataset by AGRAWAL [9], while the remaining images were taken from Ref. [8], both of which contained the star coordinates. The homogeneous dataset consists of images with greater uniformity in the luminous intensity of the stars, ensuring that no star exhibits a significantly higher brightness than the others. Forty-three images were used in this dataset.

Based on these datasets, the metrics of recall, precision, ACC, true positives (TP), true negatives (TN), false positives (FP), false negatives (FN), and the coordinate error rate generated by the model were calculated. For the latter metric, the root mean squared error (RMSE) was used, providing a quantitative measure of accuracy in locating the detected luminosity peaks.

3. Discussion and results

Table 2 presents the kernel parameters (kernel size –K_s and σ) used to generate the Gaussian kernels, which were used in the tests of the two-star identification models developed in this work. The use of a large number of kernels resulted in an increase in the execution time of the models. Through testing with a smaller number of kernels, it was observed that a set of 15 kernels, with the parameters described in **Table 2**, was sufficient to consistently identify star patterns in both 120x120 and 64x64 images. For the definition of the hyperparameters (K_s, σ, prominence, maximum intensity, and filter parameters already discussed), a validation set consisting of 10% images, separated from the datasets, was used. After this step, the final tests were conducted using the remaining images.

K_s	σ
17	3
5	3
10	3
5	5
7	7
18	5
18	13
18	11
20	13
23	13
23	15
30	15
35	15
30	25
35	25

Table 2.
Parameters used to generate the kernels for the models.

Table 3 presents the classification metrics for homogeneous and non-homogeneous datasets in terms of light intensity, using Method 1 (based on orthogonality similarity) and Method 2 (based on correlation).

In the non-homogeneous dataset, comprising 52 images with variations in noise levels and luminosity intensities, Method 1 demonstrated superior performance. It achieved an accuracy of 77%, with a recall of 73% and a precision of 1.0. In contrast, Method 2 achieved an accuracy of 59%, a recall of 54%, and a precision of 0.95. These results suggest that Method 1 is more robust in handling datasets with higher variability, while Method 2 exhibited greater sensitivity to false negatives.

For the homogeneous dataset, which consists of 43 images with uniform luminosity intensities, Method 1 achieved optimal performance, with accuracy, recall, and precision reaching all 100%.

To evaluate the accuracy of the coordinates predicted by the model in relation to the expected coordinates, the original images were labeled with the coordinates x and y corresponding to the center of the stars. By calculating the root mean square error (RMSE) of Eq. (4), it is possible to assess the precision of the predicted coordinates compared to the actual center $(\hat{x}_i, \hat{y}_i)$ of each star i, running the model for each image and analyzing the predicted coordinates (x_i, y_i). **Table 4** shows the results by indicating the exact position of the stars. Based on the achieved results, the two methods successfully identify the region of the star, with a minor failure rate in cases where the indicated points do not exactly match the central point of the star.

Dataset	Metric	Method 1	Method 2
Non-homogeneous	True positives (TP)	82	61
	True negatives (TN)	13	12
	False positives (FP)	0	3
	False negatives (FN)	30	51
	Recall	0.73	0.54
	Precision	1.0	0.95
	Accuracy (%)	77	59
Homogeneous	True positives (TP)	63	—
	True negatives (TN)	13	—
	False positives (FP)	0	—
	False negatives (FN)	0	—
	Recall	1.0	—
	Precision	1.0	—
	Accuracy (%)	100	—

[1]*The non-homogeneous data were analyzed from a set of 52 images with varying levels of noise and star luminosity intensities.*[2]*The homogeneous data consist of 43 images with uniform luminosity intensities, where no star exhibits significantly higher brightness.*

Table 3.
Evaluation results for the feature extraction methods. The table presents classification metrics for non-homogeneous and homogeneous datasets.

Metric	Method 1	Method 2
RMSE	0.517	0.169

[1]*The RMSE metric was calculated according to Eq. (4), representing the accuracy of the coordinates predicted by the model.*

Table 4.
Regression evaluation results for the feature extraction methods.

$$\mathrm{RMSE} = \left(\sqrt{\frac{(x_i - \hat{x}_i)^2 + (y_i - \hat{y}_i)^2}{2}}\right) \tag{4}$$

The post-processing stage introduced challenges that contributed to some errors in the system's final results. The use of prominence and height filters to mitigate potential noise, in certain cases, inadvertently removed relevant information from images with significant differences in luminous intensity between stars.

The tests carried out in this work, during the kernel selection phase and after their application to the model, allowed the evaluation of the effectiveness of different kernel configurations and standard deviations (σ) for object detection in blurred images, with a focus on identifying stars in astronomical images. The analysis demonstrated that the choice of standard deviation (σ) and kernel size directly affects the ability to detect objects with varying brightness and contour characteristics. A larger standard deviation generates kernels that cover broader and more diffuse areas, making them sensitive to objects with extensive brightness and poorly defined contours. Conversely, a smaller standard deviation results in more focused kernels, suitable for detecting point-like and well-defined objects. Regarding kernel size, larger kernels were observed to be more appropriate for broad structures and objects with diffuse brightness, whereas smaller kernels were more effective in identifying fine details and high-intensity objects.

Several limitations and challenges were identified during the testing. First, the exclusion of low-intensity objects emerged as a critical issue, as dimmer stars were inadvertently filtered out during the preprocessing stage when applying intensity filters. In the post-processing stage, difficulties were also observed in identifying all stars in images with significant disparities in luminous intensity among objects; in such cases, less intense stars were eliminated by prominence and height thresholds. Furthermore, many detection issues in the images were attributed to the parameters used in the post-processing stage rather than to the model itself. The correlation method showed superior performance in identifying high-intensity stars as a result of its sensitivity to contrast variations. However, it was less effective in excluding irrelevant objects, a problem that was more efficiently addressed by the orthogonality similarity method. The latter proved better suited for removing noisy objects, even when using a reduced number of kernels. Excessive kernel use, such as in tests involving up to 14,508 combinations for a 120x120 pixel image, compromised the quality of the generated maps, suggesting that a more selective kernel choice approach is recommended. Kernels with overly small dimensions can amplify noise, making the selection of a kernel set that optimally captures the characteristics of stellar images a critical factor for accurate pattern identification, as previously discussed.

A comparative analysis between orthogonality similarity and correlation indicated that the first method is more effective for star identification, prioritizing patterns and structures while ignoring intensity variations that do not affect object geometry.

This method is ideal for scenarios with variations in illumination or scale, preserving structural details. In contrast, correlation was more appropriate for objects with consistent intensity and well-defined contrasts, though it was sensitive to changes in lighting and angle, limiting its applicability in scenarios with variable luminosity.

4. Conclusions

This work developed two models for star identification in blurred images: one based on correlation and another lighter algorithm utilizing orthogonality similarity. A pre- and post-processing layer was implemented to enhance the efficiency and accuracy of the models.

In scenarios characterized by homogeneous data, the orthogonality similarity-based model demonstrated performance comparable to or even exceeding the benchmarks established in the literature. For instance, the achieved accuracy reached 100%, surpassing the results reported in related works, which range from 98% [3] to 99.27% [5], despite relying on methods that are mathematically more complex than the one proposed in this work.

For the correlation-based model tested on a non-homogeneous dataset, which presented a more challenging scenario due to variations in object sizes, noise levels, and luminosity disparities, the model achieved an accuracy close to 80%. Although this is below the metrics reported in other studies, it remains a positive outcome when considering the tested scenarios and the model's low mathematical complexity.

It is important to note that both methods developed in this work were evaluated using a small, preliminary dataset encompassing a variety of challenging scenarios, including non-homogeneous data, while the orthogonality similarity method was also assessed on a homogeneous dataset. For comparison with the literature aforementioned in the Introduction, the orthogonality similarity method demonstrated superior performance in preliminary tests, as presented in the metrics section. Homogeneous datasets were employed to align with scenarios addressed in reference studies, where stars typically exhibit similar sizes and luminous intensities. However, the images used in these studies differ significantly from those in the present work due to variations in data sources. Although prior research often utilized high-resolution images with well-controlled noise levels, this work incorporated diverse configurations that introduced additional challenges: images with single stars, stellar clusters with overlapping light, varying levels of blur, and stars positioned near image edges, complicating precise identification. These challenging visual elements highlight the need for robust algorithms capable of distinguishing real stars from noise artifacts, particularly in the peripheral regions of the image.

This work contributes to the field by advancing methodologies for addressing higher-complexity astronomical scenarios. By incorporating both homogeneous and non-homogeneous datasets, this work added diversity to the evaluation framework, demonstrating robustness and effectiveness even when working with lower-resolution images. Furthermore, the proposed models offer a balance between computational simplicity and practical applicability, showcasing their potential for real-world implementations in scenarios with varying degrees of complexity.

Future improvements include adapting the orthogonality similarity method for devices with limited storage capacity, as it presents a computationally efficient and interpretable alternative. In conclusion, the models developed in this work address significant challenges in star identification under blurred and low-resolution conditions, achieving robust results while offering potential directions for further exploration and optimization.

Acknowledgements

This work has been partially funded and supported by the Hardware Competence Center for Digital Agriculture, with financial resources from PPI HardwareBR of the MCTI grant number 056/2023, signed with EMBRAPII.

Conflict of interest

The authors declare no conflict of interest.

Author details

Aline Nunes de Souza[1†], Rodrigo Marques de Figueiredo[1,2†], Vitor Camargo Nardelli[2†] and Jean Schmith[1,2*†]

1 Unisinos University, São Leopoldo, Brazil

2 Competence Center on Digital Agriculture (EMBRAPII), SENAI Innovation Institute for Sensor Systems (ISI-SIM), São Leopoldo, Brazil

*Address all correspondence to: jean.schmith@senairs.org.br

† These authors contributed equally.

References

[1] Vijayarani DS, Sakila A, Nivetha V. Efficient blurring and de-blurring techniques for secure the document images. International Journal of Scientific and Technology Research, IJSTR©2020. 2020;**9**(03):6499

[2] Tan CY, Schulz B. A Fourier method for the determination of focus for telescopes with stars. Monthly Notices of the Royal Astronomical Society. 2022; **511**(2):2008-2020. DOI: 10.1093/mnras/stac189

[3] Zhu H, Liang B, Zhang T. A robust and fast star identification algorithm based on an ordered set of points pattern. Acta Astronautica. 2018;**148**: 327-336

[4] Zhao Y et al. Star identification algorithm based on k–l transformation and star walk formation. IEEE Sensors Journal. 2016;**16**(13):5202-5210

[5] Liang X et al. Star identification algorithm based on image normalization and zernike moments. IEEE Access. 2020;**8**:29228-29237

[6] Zou Y et al. Segmenting star images with complex backgrounds based on correlation between objects and 1d Gaussian morphology. Applied Sciences. 2021. Available from: https://www.mdpi.com/2076-3417/11/9/3763;**11**(9):3763. [Accessed: November 11, 2024]

[7] Liu M et al. Star identification based on multilayer voting algorithm for star sensors. Sensors. 2021. Available from: https://www.mdpi.com/1424-8220/21/9/3084;**21**(9):3084. [Accessed: June 13, 2024]

[8] Abdurro'uf et al. The seventeenth data release of the Sloan digital sky surveys: Complete release of manga, mastar, and apogee-2 data. The Astrophysical Journal Supplement Series, The American Astronomical Society. 2022;**259**(2):35

[9] Agrawal D. Star-Galaxy Classification Data; 2021. Available from: https://www.kaggle.com/datasets/divyansh22/dummy-astronomy-data/data [Accessed: June 21, 2024]

[10] Wang H et al. Gaussian analytic centroiding method of star image of star tracker. Advances in Space Research. 2015. Available from: https://www.sciencedirect.com/science/article/pii/S0273117715006110;**56**(10):2196-2205 [Accessed: July 21, 2024]

Chapter 4

Fire Detection Using Image Processing and Machine Learning

Anderson Felipe Weschenfelder, Jéssica Diehl, Renan Santos dos Santos, Rodrigo Marques de Figueiredo, Vitor Camargo Nardelli and Jean Schmith

Abstract

Fire detection plays a crucial role in safeguarding agricultural lands, pastures, and forested areas from catastrophic damage. The presence of diverse textures, colors, and shapes in these environments makes fire detection challenging, with many existing methods prone to high false-positive rates. In this work, we propose a two-stage fire detection method utilizing k-Nearest Neighbors (kNN). First, kNN is used to classify images as containing fire or not, achieving an accuracy of 81.57%. For images classified as containing fire, they are further divided into sub-images, and additional features are extracted. kNN is then applied again to classify these sub-images to localize the fire regions, achieving an accuracy of 84.81%. This approach is particularly effective in agribusiness contexts, where the shades of green dominate the landscape and the presence of orange and red is strongly correlated with fire. The proposed method offers a robust solution for the early detection and location of fires on agricultural land, supporting the protection of crops, pastures, and forest borders while enabling a rapid response to minimize economic and environmental losses.

Keywords: fire detection, feature extraction, digital image processing, machine learning, kNN

1. Introduction

Forest fire detection through image analysis is a rapidly advancing research field [1]. With advances in remote sensing technology, it has become possible to utilize image processing algorithms for early-stage fire identification, leading to faster and more effective response strategies. Climate change plays a critical role in increasing the risk of forest fires, as hot and dry weather conditions create a conducive environment for rapid fire spread [1, 2].

Remote fire detection offers significant advantages, such as broader coverage and the ability to monitor inaccessible areas. Surveillance cameras, drones, and satellites can be flexibly deployed for image-based detection, tailored to specific regional requirements [1]. This enables a prompt response and helps prevent infrastructure damage, protect human communities, and preserve wildlife. Furthermore, effective fire prevention at an early stage can reduce greenhouse gas emissions, such as carbon dioxide, thus mitigating global warming [2]. Thus, remote fire detection in forests is an ever-evolving field, which is harnessed by remote sensing technology. The link between climate change and forest fires emphasizes the importance of early detection in risk mitigation. By allowing for broad coverage and rapid response, remote detection contributes to damage prevention and reduction of greenhouse gas emissions.

Forest fire detection through image analysis is a promising technique that offers significant advantages, such as early and accurate fire detection, flexibility in system installation, and the ability to effectively monitor large areas [3]. This approach is essential in addressing the growing challenges posed by climate change and in minimizing the resulting damage.

In this context, Chen et al. [4] proposed an innovative approach for early fire detection using a combination of video and color information. By employing an Red, Green and Blue (RGB) model based on chromatic measurements and clutter, their method aimed to extract fire pixels in videos, enabling the identification of fire occurrences. This approach demonstrates the potential of using image analysis techniques to improve forest fire detection systems and improve overall fire management strategies.

In addition to its potential applications in forest fire detection, the fire detection method discussed in that work can be adapted to monitor fire outbreaks in cultivars, offering significant benefits for the agribusiness sector. The ability to rapidly detect fires in agricultural areas is crucial, as it can help prevent the widespread destruction of crops, protect valuable resources, and minimize economic losses. Agricultural fields, especially in regions prone to drought and heat, are highly vulnerable to fire outbreaks, which can be exacerbated by climate change.

Based on this, 1D and 2D wavelet transform methods were developed to detect fire motion regions by integrating information on color and temporal variation [5]. According to Chino et al. [6], most fire detection algorithms were developed for videos with obvious limitations. To address this problem, a novel fire detection method called BowFire was proposed. This method combines color features with superpixel texture discrimination to detect fire in still images.

In addition, Sharma et al. [7] employed optical images and fine-tuned two pre-trained convolutional neural networks (CNNs), using VGG16 and ResNet50 architectures, to distinguish between images containing fire and those that do not. Furthermore, Dunnings and Breckon [8] used a CNN with a lower complexity architecture, applying superpixels to reduce computational requirements. Despite the good results, CNN training demands large datasets, large computational capacity, and much memory consumption [9].

In this proposed work, the fire detection method developed by Chino et al. [6], was used as a reference for the comparison of results. The BowFire method, which combines color features with superpixel texture discrimination, was considered an established and widely recognized approach for fire detection in still images. Using this method as a benchmark allowed the evaluation of the effectiveness and performance of our proposed new method, providing valuable insights into its detection capabilities and potential improvements compared to the results achieved by the existing method. This comparative approach contributes to a comprehensive analysis and a more accurate assessment of the innovation and efficiency of the new fire detection method.

After initial research on fire detection methods and common approaches to solving this task, we decided to test how well a simplistic and faster method would perform. In this work, we focus on the k-Nearest Neighbors (kNN) algorithm in the image domain.

The method applied consisted of two stages. The first stage aimed to detect the presence of fire in an image as a whole, using the kNN algorithm. The second stage was a sequential step applied only to images in which fire was detected during the first stage. Upon detection of the fire, the second stage aims to define where the fire is located in the image. For this task, the images were divided into smaller regions, and then, kNN was applied to classify each region in particular. The regions flagged in the second stage, therefore, indicate an approximated position of the fire when projected on the original image. All of this is described in detail in the methodology section.

2. Methodology

The proposed method is presented in the diagram depicted in **Figure 1**. It involves two stages of classification. In the first stage, kNN is used to classify an image as either fire or no-fire. If the image is classified as fire, it is divided into smaller parts. For each of these parts, various features are calculated, and kNN is applied once again to determine whether each part contains fire or not. Based on these results, the original image is reconstructed, highlighting areas where the fire is present.

2.1 Dataset

For training, testing, and evaluating our proposed method, we used the dataset from Khan and Hassan [10] with 1900 images. The dataset is divided into fire and no-fire, with 950 images in each class. The images are colored, with a resolution of 250 × 250 pixels, and are already split into 80% for training and 20% for testing machine learning algorithms.

2.2 Applying kNN to full image dimensions

The main idea was to first classify the images as containing fire or not. Therefore, the dataset was processed using Python programming and the OpenCV library [11]. Each image was decomposed as a vector, and each RGB channel was aligned as a single vector; thus, each image vector has a size of 187,500 (250 × 250 × 3). At the last position of the vector, the information containing the fire/no-fire information was added as 1 or 0, respectively, as shown in the example in **Figure 2**. This process was applied to all images in the fire and no-fire dataset; each one was finally composed of 760 rows (number of images in the training dataset) and 187,500 columns plus the information of fire/no-fire.

Both datasets (fire and no-fire) were concatenated and resulted in the training dataset. It was not necessary to divide these data between training and testing, as the original data already carried a reserved folder with images specifically for testing.

kNN was applied to the training dataset using the Scikit Learn library [12]. Finally, the best results were achieved with 50 neighbors. To evaluate the algorithm, the trained

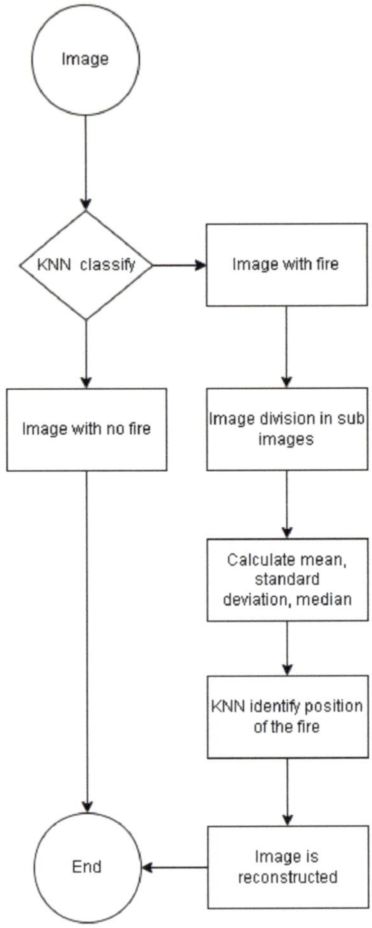

Figure 1.
Block diagram of the proposed method for fire detection.

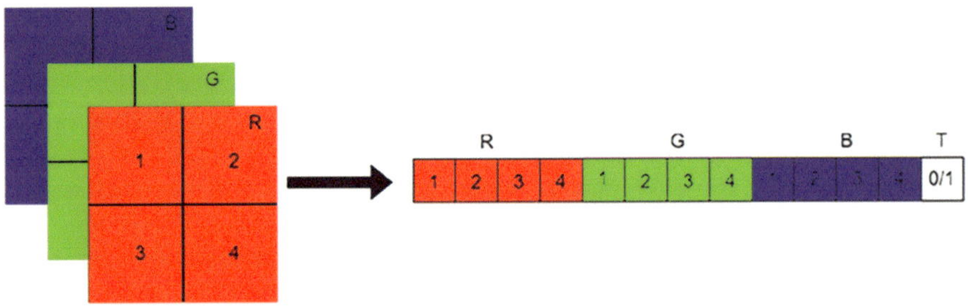

Figure 2.
Example of an RGB image of resolution 2 × 2 pixels rearranged in a vector of 13 positions, including the fire/no-fire information in the slot "T."

model was tested using the test dataset. We used the accuracy, precision, recall, and F1-score as evaluation metrics given by the Eqs. (1), (2), (3), and (4), respectively, computed from the confusion matrix, namely true positives (TP), true negatives (TN), false positives (FP), and false negatives (FN) of the classification.

$$accuracy = \frac{TP + TN}{TP + TN + FP + FN}. \quad (1)$$

$$precision = \frac{TP}{TP + FP}. \quad (2)$$

$$recall = \frac{TP}{TP + FN}. \quad (3)$$

$$F1score = 2\frac{precision.recall}{precision + recall}. \quad (4)$$

2.3 Image division

After the initial classification of an image as a fire, it is further processed by another algorithm that aims to identify and visualize the specific locations of the fire within the image. The core concept is to divide the original image into smaller parts (sub-images). Therefore, various features and patterns can be extracted to distinguish between images that contain fire and those that do not. Subsequently, these extracted features are used to reconstruct the original image, highlighting the precise areas where the fire is present.

In the dataset used for this study, the images initially have a resolution of 250 × 250 × 3 pixels and consist of three color channels: red, green, and blue. The division process involves splitting each color channel matrix into smaller matrices with dimensions of 25 × 25 × 3 pixels. We tested different resolutions for the sub-images, and the best results were achieved with 25 × 25 × 3 pixels. Consequently, for a color image, a total of 100 sub-images are generated, each retaining its respective color information. **Figure 3** illustrates the outcome of this division process, with each sub-image uniquely identified by a number ranging from 0 to 99, providing a clear visualization of the resulting sub-images.

To establish correlations between fire and no-fire instances, a manual classification process was used to label the presence or absence of flame-fire in different image segments; for each to be considered fire, the flame must cover over 50% of the sub-image area, and thus, the smoke is not considered fire in this work. A set of 20 images was randomly chosen from the dataset, and each image was subdivided into 100 smaller sub-images. The same approach as described in Section 2.2 was used to organize the sub-images into a comprehensive dataset. This dataset comprised 100 rows and 1875 columns, corresponding to 25 × 25 × 3 dimensions for each image.

A comprehensive analysis was performed on a dataset consisting of sub-images containing flame fires. The goal was to identify patterns and characteristics that could differentiate these sub-images. To achieve this, various statistical features were computed for each sub-image, and we selected the most representative ones to distinguish fire into the sub-images. The selected features included the global standard deviation (σ Global) representing the overall variation in the pixel values throughout the image. Furthermore, the mean (μ Red), median (Median Red), mode (Mode Red), and harmonic mean (H Red) of the red channel.

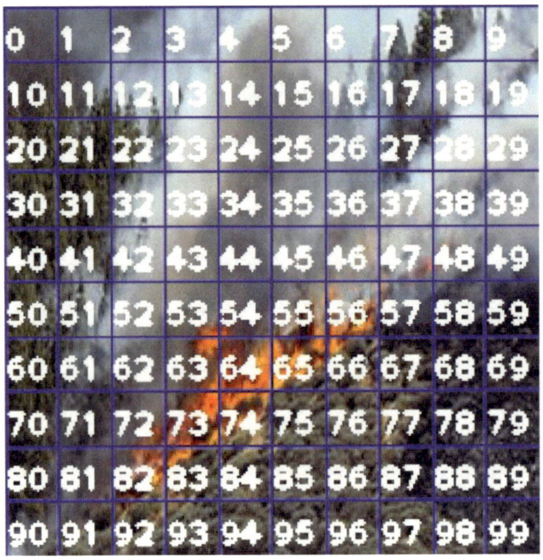

Figure 3.
Example of an image divided into 100 small parts (sub-images).

The computation of these statistical measures provided valuable insights into the characteristics and variations exhibited by the sub-images. Visualizing the distribution of these characteristics through a box plot, as shown in **Figure 4**, allowed the identification of potential patterns and trends associated with flames. One might note the good separation between fire and no-fire box plots for the selected features. It was observed that the features of the blue and green channels displayed values in the same order for both fire and non-fire situations, rendering them less effective for distinguishing between the two. Consequently, these characteristics were ignored in the further analysis. On the other

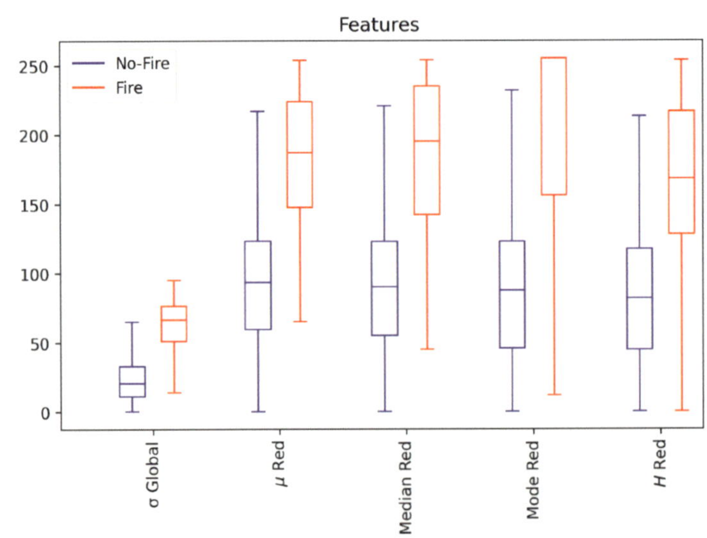

Figure 4.
Box plot of the feature distribution values of fire and no-fire sub-images.

hand, the features derived from the red channel demonstrated superior performance, as the color red is particularly prominent in flame fires.

The kNN algorithm was employed to compute the computed selected features for each sub-image. The training and testing methodology remained consistent with the previous approach. Each sub-image was classified as either fire or no-fire manually.

A new training dataset was generated, covering all computed features for the 20 images and their corresponding sub-images. A column was once again appended to indicate the presence or absence of fire in each sub-image. The kNN classifier was trained using this dataset.

For testing purposes, an additional set of 53 images was used. A test dataset was constructed, consisting of the divided sub-images and their respective classifications, the same way as images for training. The features were then calculated and inputted into the kNN model with 50 neighbors. The predicted values from the kNN model were subsequently compared with the original values, following the same evaluation metrics as described earlier.

3. Results and discussion

In this study, kNN was applied in two different classifications using the same dataset. Each classification transformed and preprocessed the dataset in different ways. The first classification focused on detecting the presence of fire and leveraged each image in the dataset, while the second classification task divided the image into smaller sections to specify the region containing the fire.

The performance of the kNN model was evaluated by comparing its predictions to the corresponding target values. An important aspect of the evaluation is to be consistent in measuring true positives for the image as a whole, meaning that detecting the presence of fire is a positive event. The absence of fire is a negative event and was not the focus of the task.

The dataset in Ref. [10] was carefully selected to ensure that both positive and negative events were balanced, so when the model was evaluated, the results were not skewed toward one of the events.

3.1 Results of applying kNN to full image dimension

Table 1 presents the performance evaluation of the full image classification, which aimed to classify whether the image contains fire or not. The table presents various performance metrics, including accuracy, precision, recall, and F1-score.

To perform the classification, the code takes 84 ms, on a computer with an Intel Core I7-4500 U CPU @ 1.8 GHZ, considering the time to encode the fire, encode the no-fire images, concatenate the datasets, and predict the fire in each image.

Accuracy	81.57%
Recall	91.57%
Precision	76.31%
F1-score	83.25%

Table 1.
Performance evaluation of full image classification into fire and no-fire.

Although the kNN method is a simple and straightforward approach by Refs. [13, 14], this research shows that it can produce high accuracy as an initial detection mechanism. Moreover, we notice a high recall that supports the effectiveness of the intended task, as it maximizes the detection of the positive event.

3.2 Results of sub-image classification

To enhance the performance of the flame-fire detection system, we conducted experiments with various classification algorithms, including support vector machine (SVM), naive Bayes, random forest, and kNN. Among these algorithms, kNN demonstrated the best performance, achieving an accuracy of 84.81% as presented in **Table 2**. These outcomes indicate the effectiveness of the chosen features with the proposed methodology in accurately classifying images and detecting flame-fire instances and also point to the kNN as a good choice of the machine learning algorithm.

Table 3 presents the performance evaluation of the sub-image classification, which aimed to accurately identify the location of the fire within an image. This table also presents various performance metrics, including accuracy, precision, recall, and F1-score, providing information on the effectiveness of the classification model in detecting the presence of fire in different parts of the image.

After training the kNN model and using it to predict and reconstruct the original image, the process of highlighting the specific areas where the fire is present takes 517 ms. It is important to note that this execution time can be further reduced by using a more powerful computer or further optimization of the code for faster execution.

It was proposed by Chino et al. [6] a flame-fire detection method that uses two classifiers: color-based classification with Naive-Bayes and texture-based classification with kNN. They conducted a comparative analysis of their results with those of Celik and Demirel [15]. Rudz et al. [16] considered two scenarios: one with the inclusion of texture information and one without it. Muhammad et al. [3] proposed the use of a convolutional neural network (CNN) to detect fire, and their results were also compared to other works.

Method	Accuracy	Recall	Precision	F1-score
kNN	84.81%	81.80%	79.27%	80.52%
SVM	84.60%	82.74%	78.35%	80.48%
Random forest	83.49%	88.34%	73.79%	80.42%
Naive Bayes	80.92%	90.95%	69.10%	78.54%

Table 2.
Performance comparison with different machine learning methods.

Accuracy	84.81%
Recall	81.80%
Precision	79.27%
F1-score	80.52%

Table 3.
Performance evaluation of sub-image classification into fire and no-fire.

Work	Precision	Recall	F1-score
Our proposed method	79.27%	81.80%	80.52%
Muhammad et al. [3]	86.00%	89.00%	88.00%
Chino et al. [6]	40–60%	60–80%	60–70%
Celik et al. [15]	40–60%	50–60%	50–60%
Rudz et al. [16]	60–70%	40–50%	50–60%

Table 4.
Comparison of our method with state-of-the-art methodologies.

When comparing their best results with our proposed study, as shown in **Table 4**, it becomes evident that our method achieves good performance, only Muhammad et al. [3] with CNN achieved higher results; however, it requires a powerful computer and a larger database for training. Li and Zhao [17], who also used CNN, achieved an accuracy of 99.62%, although it also required 29,180 images and a high-performance computer.

Although the datasets of related works are not the same as ours, the proposed method demonstrates competitive performance compared to state-of-the-art techniques while maintaining efficiency and accuracy. This highlights the potential of our approach for fire detection, especially in the Khan and Hassan [10] dataset.

Testing our proposed method with different images yielded promising results. **Figures 5** and **6** illustrate where the fire was correctly detected, highlighting only the parts where the flame-fire is present. However, in cases where the flame-fire is smaller, the detection response is not as accurate as presented in **Figure 7**. One explanation for these errors is that the sub-image resolution is much higher than the fire region. Nonetheless, the method is still able to identify the general location of the fire origin.

The fire detection system exhibited a high level of accuracy in various scenarios, with the exception of certain instances during sunset in **Figure 8**. During this particular lighting condition, the system encountered some challenges and produced errors

(a) (b)

Figure 5.
Example of correct detection of fire region in the image. (a) Original image; (b) Our method output.

(a) (b)

Figure 6.
Another example of correct detection of a fire region in the image. (a) Original image; (b) Our method output.

(a) (b)

Figure 7.
Example of detection of fire region in image with missing parts. (a) Original image; (b) Our method output.

(a) (b)

Figure 8.
Example of incorrect detection of fire region in sunset image. (a) Original image; (b) Our method output.

in fire detection. The errors can be attributed to the changes in color and illumination that occur during sunset, which affect the performance of the fire detection algorithm. Further improvements are required to enhance the system's performance during such challenging lighting conditions.

4. Conclusion

As kNN is an algorithm based on simplistic data segmentation in hyper-dimensional space, it is expected that the classification of fire is based on the pixel values, mainly on the red channel of the RGB. This behavior introduces some limitations to the detection capabilities. Among these limitations, it recognizes that very small regions of fire in the image are not well detected by the kNN, which is somewhat improved by the second classification that breaks down the image into smaller regions. Another recognized limitation is that images with the absence of fire, taken during sunrises or sunsets, or images taken during the fall season, introduce a data pattern with higher values on the red channel. This pattern leads to false positives during the detection, as the kNN algorithm understands that images dominated by orange and red colors are similar to fire.

Still, even with recognized limitations, the method proves to be accurate, especially in forests dominated by shades of green and where the presence of orange and red is almost always indicated by the presence of fire. In terms of computational cost during training and inference time, kNN shows good results and is suitable for real-time applications with a simpler approach in comparison to more complex and elaborate neural networks such as CNNs, using VGG16 and ResNet50 architectures. Future directions point to the application of our proposed method in surveillance cameras and in testing in real-world applications.

Acknowledgements

This work has been partially funded and supported by the Hardware Competence Center for Digital Agriculture, with financial resources from PPI HardwareBR of the MCTI grant number 056/2023, signed with EMBRAPII.

Conflict of interest

The authors declare no conflict of interest.

Author details

Anderson Felipe Weschenfelder[1†], Jéssica Diehl[1†], Renan Santos dos Santos[1†], Rodrigo Marques de Figueiredo[1,2†], Vitor Camargo Nardelli[2†] and Jean Schmith[1,2*†]

1 Unisinos University, São Leopoldo, Brazil

2 Competence Center on Digital Agriculture (EMBRAPII), SENAI Innovation Institute for Sensor Systems (ISI-SIM), São Leopoldo, Brazil

*Address all correspondence to: jean.schmith@senairs.org.br

†These authors contributed equally.

References

[1] Xu R, Lin H, Lu K, Cao L, Liu Y. A forest fire detection system based on ensemble learning. Forests. 2021;**12**(2):217

[2] Jones MW, Abatzoglou JT, Veraverbeke S, Andela N, Lasslop G, Forkel M, et al. Global and regional trends and drivers of fire under climate change. Reviews of Geophysics. 2022;**60**(3):e2020RG000726

[3] Muhammad K, Ahmad J, Mehmood I, Rho S, Baik SW. Convolutional neural networks based fire detection in surveillance videos. IEEE Access. 2018;**6**:18174-18183

[4] Chen T-H, Wu P-H, Chiou Y-C. An early fire-detection method based on image processing. In: 2004 International Conference on Image Processing, 2004, ICIP'04. Singapore: IEEE; 2004. pp. 1707-1710

[5] Töreyin BU, Dedeoğlu Y, Güdükbay U, Cetin AE. Computer vision based method for real-time fire and flame detection. Pattern Recognition Letters. 2006;**27**(1):49-58

[6] Chino DYT, Avalhais LPS, Rodrigues JF, Traina AJM. Bowfire: Detection of fire in still images by integrating pixel color and texture analysis. In: 2015 28th SIBGRAPI Conference on Graphics, Patterns and Images. Salvador, Brazil: IEEE; 2015. pp. 95-102

[7] Sharma J, Granmo O-C, Goodwin M, Fidje JT. Deep convolutional neural networks for fire detection in images. In: Engineering Applications of Neural Networks: 18th International Conference. Athens, Greece: Springer; 2017. pp. 183-193

[8] Dunnings AJ, Breckon TP. Experimentally def ined convolutional neural network architecture variants for non-temporal real-time fire detection. In: 2018 25th IEEE International Conference on Image Processing (ICIP). Athens, Greece: IEEE; 2018. pp. 1558-1562

[9] Ponti MA, dos Santos FP, Ribeiro LSF, Cavallari GB. Training Deep Networks from Zero to Hero: Avoiding pitfalls and going beyond. In: 2021 34th SIBGRAPI Conference on Graphics, Patterns and Images (SIBGRAPI). Gramado, Brazil: IEEE; 2021. pp. 9-16

[10] Khan A, Hassan B. Dataset for Forest Fire Detection. 2020. Available from: https://data.mendeley.com/datasets/gjmr63rz2r/1

[11] Bradski G. The OpenCV library. Dr. Dobb's Journal of Software Tools. 2000;**25**:120-123

[12] Pedregosa F, Varoquaux G, Gramfort A, Michel V, Thirion B, Grisel O, et al. Scikit-learn: Machine learning in python. Journal of Machine Learning Research. 2011;**12**:2825-2830

[13] Kerscher PJP, Schmith J, Martins EA, de Figueiredo RM, Keller AL. Steel type determination by spark test image processing with machine learning. Measurement. 2022;**187**:110361

[14] Kinoshita NYK, Schmith J, Martins EA, de Figueiredo RM. A method for identifying vegetation under distribution power lines by remote sensing. Journal of Control, Automation and Electrical Systems. 2023;**34**(6):1284-1293

[15] Celik T, Demirel H. Fire detection in video sequences using a generic

color model. Fire Safety Journal. 2009;**44**(2):147-158

[16] Rudz S, Chetehouna K, Hafiane A, Laurent H, Séro-Guillaume O. Investigation of a novel image segmentation method dedicated to forest fire applications. Measurement Science and Technology. 2013;**24**:075403

[17] Li P, Zhao W. Image fire detection algorithms based on convolutional neural networks. Case Studies in Thermal Engineering. 2020;**19**:100625

Chapter 5

Object Detection Algorithms for Digital Imaging Applications: A Review

Rotimi-Williams Bello, Pius A. Owolawi, Etienne A. van Wyk and Chunling Tu

Abstract

Object detection is a major branch and fundamental task in computer vision, aiming to localize, identify and classify even the smallest objects of interest in images. Features can be extracted efficiently by deep convolutional neural networks (CNNs) as the backbone for real-time or near real-time object detection performance than the hand-crafted-based traditional methods. In the past few years, the advent of transformer-based models with robust self-attention mechanisms has not only raised object detection performance to a higher level but has also enabled it to produce excellent results. Many object detection tasks in the real world require that 3D information about the object be obtained, thus strengthening active research in 3D object detection. However, the algorithms for detecting 3D objects are not easy to propagate in real-world applications due to many factors, making reconstruction of 2D object detection algorithms to 3D object detection algorithms the suitable alternative. Therefore, we review the evolution of 2D object detection algorithms for digital imaging applications, focusing on their developments, models, applications, datasets, evaluation metrics, strengths and weaknesses, for better understanding of their landmarks and contributions to the advancement of the field.

Keywords: algorithms, convolutional neural networks, digital imaging, models, object detection and classification

1. Introduction

Object detection is a fundamental task in the field of computer vision (CV) that primarily seeks how objects can be localized, identified, and classified into predefined categories of interest in images. Many downstream vision problems such as object tracking and monitoring depend on object detection algorithms, making it paramount to intensify research on how object detection models can be enhanced to produce effective and high-performance results. In the past few years, object detection has evolved significantly following the remarkable advancements in CV [1]. There are three stages included in the traditional algorithms developed for object detection, namely region proposal stage (for locating potential objects), feature extraction stage, and classification stage.

The sliding window (of different sizes) method was the popular approach employed by the region proposal stage to extract regions (objects) of interest (RoI) from the input image, and this is to ensure that candidate regions are obtained by multi-iterations irrespective of the different sizes that each target object may possess. Talking about the second stage, in its incipient form, object detection tasks relied on the features extracted by hand-crafted-based traditional methods, such as Local Binary Pattern (LBP) [2], Scale-Invariant Feature Transform (SIFT) [3, 4], Histogram of Oriented Gradients (HOG) [5]. The features extracted by hand-crafted-based traditional methods are utilized in the last stage for the object classification and regression of bounding boxes.

However, there are a lot of significant flaws in these traditional algorithms and other traditional object detection models like Viola-Jones [6, 7], such as low-processing speed and accuracy, generalization issues and high operating cost, which deep convolutional neural networks (CNNs) have steadily replaced. The feat that CNNs perform with the advent of AlexNet [8] in 2012, R-CNN (Region-Based CNN) [9] in 2014 and R-CNN series has led to a paradigm shift in deep learning-based methods for object detection and CV. The dominance of deep learning has been spread to various areas of object detection, leading to tremendous growth of various methods developed for object detection, motivated by high-processing speed and accuracy, capability for generalization, large-scale datasets and progress made in other CV tasks.

The frameworks of modern object detection models are primarily classified into two-stage object detection models, one-stage object detection models, and transformer-based object detection models, as shown in **Figure 1**.

The two-stage object detection models are typified by the R-CNN series, such as Faster R-CNN [10], Mask R-CNN [11], performing object detection in two steps, which is by first-generating region proposals before classifying and refining those regions. The one-stage object detection models, such as You Only Look Once (YOLO) [12], Single Shot MultiBox Detector (SSD) [13], GluonCV [14], perform detection

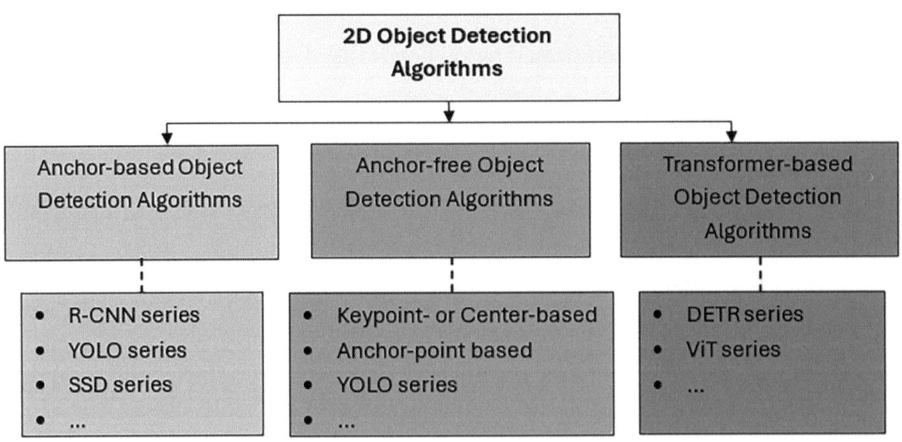

Figure 1.
Classifications of 2D object detection algorithms for digital imaging applications.

directly from the input image by directly predicting bounding boxes and class labels for objects within an image in a single pass, prioritizing efficiency and speed over potential accuracy prioritized by the two-stage object detection models.

Transformer-based object detection models are the most recent object detection models that have DEtection Transformer (DETR) [15] as a typical example. Global context across the entire image is captured by the detection models by utmost reliance on the self-attention mechanism in Transformer architecture [16]. The network's backbone (AlexNet, ResNet [17], Swin Transformer [18], etc.), the network's neck (Feature Pyramid Network (FPN) [19], YOLOF [20], DETR, etc.) and the network's head (YOLO, SSD, Region-based Fully Convolutional Networks (R-FCN) [21], etc.) are the three parts into which network is divided when processing images that are fed into the network (**Figure 2**). The three parts serve different purposes, starting from the backbone, through which feature extraction is performed, next is the head, through which the extracted features are merged and refined, and the head, through which the object's classes and location coordinates are predicted.

Anchor-based object detection models, such as those previously mentioned, are mostly developed in series with the aim of improving on those series to leverage the processing speed and performance accuracy of the detection models, making them superior to the existing traditional algorithms. The development of anchor-free object detection models eliminated the limitations in using the preset anchor boxes by introducing key points or anchor points for bounding box regression. Many proposals have been presented on using Transformer as either the network's neck or backbone, and the proposals were motivated by the wide acceptance of Transformer's application to the field of CV with Vision Transformer (ViT) [22], producing promising results.

Many object detection tasks in the real world require that 3D information about the object be obtained, thus strengthening active research in 3D object detection. However, the algorithms for detecting 3D objects in 3D space are not easy to propagate in real-world applications due to many factors, making reconstruction of 2D object detection algorithms to 3D object detection algorithms the suitable alternative for 3D object detection in 3D space. Therefore, we review the evolution of 2D object detection algorithms for digital imaging applications, focusing on their developments,

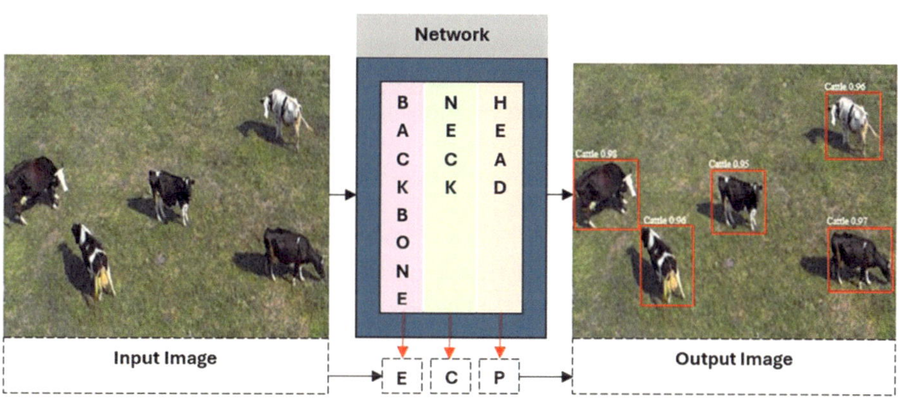

Figure 2.
The three parts into which the network is divided when processing images that are fed into the network are the network's Backbone (AlexNet, ResNet, Swim Transformer, etc.), the network's Neck (FPN, YOLOF, DETR, etc.) and the network's Head (YOLO, SSD, R-FCN, etc.). The Backbone is for features Extraction (E), the Neck is for Combining (C) the extracted features and the Head is for generating the final Prediction (P) based on the features extracted by the Backbone and Neck layers.

models, applications, datasets, evaluation metrics, strengths and weaknesses, for better understanding of their landmarks and contributions to the advancement of the field.

To achieve this, we collected and analyzed recent high-quality CV-centered research papers from top journals, book chapters and conferences that focused on algorithms for detecting digital imaging objects. The contributions of this chapter are as follows:

1. In this chapter, an in-depth review was carried out on the 2D object detection, covering detailed analysis of the main paradigms, starting from the hand-crafted-based traditional methods to deep learning-based modern methods.

2. Each paradigm was analyzed to cover its development overview, milestones and key innovative contributions toward achieving the modern methods for detecting 2D objects.

3. This chapter discusses datasets, and metrics for evaluating object detection, before delineating future directions in this rapidly evolving field.

4. This chapter divides the algorithms of the 2D object detection into anchor-based detection models, anchor-free detection models and Transformer-based detection models, covering detailed analysis of their contributions.

2. Algorithms for detecting 2D objects

2.1 Hand-crafted-based traditional methods

Object detection, in its early period, relied heavily on the features extracted by the hand-crafted-based traditional methods, and these methods involved extracting discriminative details from target objects using hand-crafted characteristics. Feature extraction methods that belong to this category are HOG, SIFT, Integral Image [6, 7], etc. The extracted features were merged and processed by the classification models of that time, like Support Vector Machines (SVM) [23], SVM series such as LSVM (Lagrangian SVM) [24], SO-SVM [25, 26] and AdaBoost [27]. The object detection model invented by Viola-Jones (named Viola-Jones detector), the HOG-based object detection models and the Deformable Part-based Model (DPM) [28] are good examples that contributed remarkably to object detection during their time.

Moreover, other object detection models that existed during that period relied on SIFT descriptors or on their series, such as color-SIFT descriptors [29] and PCA-SIFT descriptors [30]. Likewise, other object detection models of that period employed more feature extractors, such as Haar-like wavelets [31], Hough transform [32]. Although the hand-crafted-based traditional methods significantly addressed the challenges confronting object detection during their time, low-processing speed and complexity in computation are their notable limitations. To address these limitations, deep learning-based object detection models were developed and categorized into three, namely anchor-based object detection models (use CNNs for the detection process), anchor-free object detection models (use CNNs for the detection process) and Transformer-based object detection models (use Transformer architecture).

Although there was so much technological advancement in the models of anchor-based object detection, these models were undermined by their ineffectiveness in multiscale detection tasks. Likewise, the models of anchor-free object detection gained wide acceptance due to their capability for direct prediction of object bounding boxes from the pixels of the images exclusive of the pre-defined anchor boxes, resulting in a better and more robust approach than the earlier object detection methods. Transformer-based object detection models possess similar qualities as anchor-free object detection models, representing recent state-of-the-art object detection models that use Transformer architecture for extraction of object's features from images and their combination.

2.2 Anchor-based object detection algorithms

Anchor-based object detection models are types of CV model for object detection that rely on pre-defined bounding boxes (also known as anchor boxes) for object's localization and classification within an image, effectively acting as points of reference to facilitate the model's capabilities in identifying potential object areas by fine-tuning their position and size to match the actual objects in the image. These object detection algorithms are divided into R-CNN series, YOLO series and SSD series.

2.2.1 R-CNN series anchor-based object detection algorithms

R-CNN (with its series) is the most influential object detection model in CV that allows many candidate bounding boxes of different sizes to be generated for the input image by selective search algorithm [33]. The generated bounding boxes are input into the CNNs for feature extraction prior to their classification by SVM and location fine-tuning by linear regressor. The approach of pre-training the model on many datasets and fine-tuning it on a few numbers of datasets significantly mitigates the data limitation problems. As shown in **Figure 3**, the advent of R-CNN brought about using CNNs for extraction of features from images, making it a better replacement to traditional algorithms in terms of high-processing speed, accuracy, generalization ability and simple computation.

However, R-CNN lacks efficiency and thus affects the training and inference phases. Moreover, the approach of dividing the detection process into four stages negatively affected both the processing speed of the model training and the processing

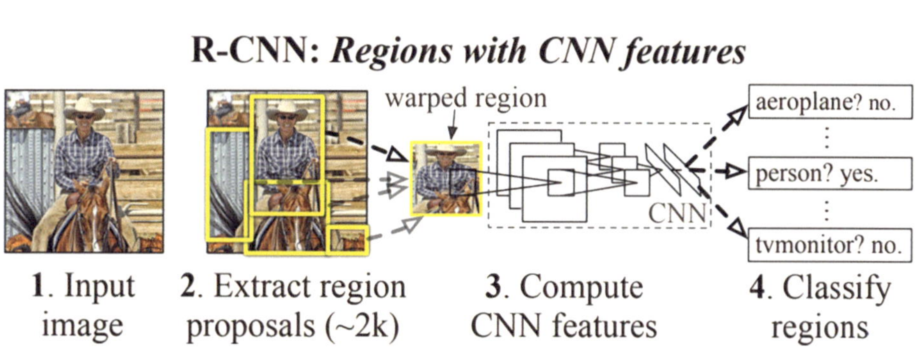

Figure 3.
Overview of R-CNN object detection model [9]. The model (1) accepts input image, (2) extracts from the image, approximately 2000 bottom-up region proposals, (3) computes CNN features for each proposal and then (4) classifies the regions.

resources. SPPNet [34] was proposed as a solution to the above-mentioned limitations of R-CNN by adding a layer of Spatial Pyramid Pooling (SPP) [35] prior to the fully connected (FC) layer. The size of the candidate box can be integrated by SPPNet with just one convolution operation on the image, thereby improving the model's speed. Moreover, the training of multi-size images is permitted by SPPNet and the robustness of multi-level pooling to object deformation is also improved.

Although the detection speed of SPPNet is impressive, there is still a need for multi-stage training. In addition, the overall accuracy may be affected by SPPNet due to fine-tuning only the FC layer while fixing the convolutional layers, to simplify the training and increase the speed of the network.

Fast R-CNN [36] was proposed to fix the limitations of R-CNN and SPPNet while maintaining and improving their strengths. A RoI pooling layer is utilized by Fast R-CNN for the combination of extracted image features, making them suitable as input into the FC layer, whose role is to classify and regress. Fast R-CNN has several advantages compared to R-CNN, and these advantages include faster inference and single-stage training by feature sharing and a multi-task loss. However, the speed at which it detects an object in real-time is low due to the process involved in object proposal and large-scale datasets used, motivating Faster R-CNN development as a solution to Fast R-CNN's limitations by merging proposal generation in the network.

Faster R-CNN comprises two modules, namely (1) a deep FCN [37], otherwise known as Region Proposal Network (RPN), by which regions are proposed, and (2) Fast R-CNN, by which the proposed regions are utilized. Simply put, the anchor candidates are generated by the RPN, which has some common characteristics with the detection network, before the RoI is mapped to a fixed-size feature map by the RoI pooling layer, leading to the generation of predicted results by the FC layer. This network arrangement by Faster R-CNN makes object detection an end-to-end 3-in-1 process by using same CNNs to generate candidates, extract features, and classify and regress bounding boxes, making its speed faster than the speed of Fast-R-CNN in real-time object detection.

However, there are some learnable layers that are not made convolutional with Faster R-CNN. To address the drawbacks in Faster R-CNN, R-FCN [21] was proposed with some impressive approaches. R-FCN is designed as FCN without FC layer and ends with a position-sensitive RoI pooling layer for score maps aggregation and scores generation for each RoI. The end-to-end training of R-FCN enables the RoI layer in guiding the last layer of the CNNs to learn position-sensitive score maps, solving CNN's translation invariance. The design of R-FCN enables fast training and inference, and region-wise computation at little cost. However, R-FCN is costly due to the many channels in the feature maps before RoI pooling [38].

Successive works that are based on R-FCN include Deformable R-FCN (D-RFCN) [39] and Light Head R-CNN (LH R-CNN) [40] that address the cost in R-FCN by introducing separable convolution. The RoI pooling layer in R-FCN was replaced with the RoI alignment layer in Mask R-CNN. Mask R-CNN is an extension of Faster R-CNN, developed for instance segmentation [1, 41–43]. Mask R-CNN combines residual network (ResNet) and FPN as its backbone network, in which the FPN framework was designed for the improvement of multi-scale object detection with computation efficiency. For a long period of time, FPN has been generalized to include Parallel Feature Pyramid Network (PFPNet) [44] and Attention Aggregation FPN (A2-FPN) [45]. The mapping of RoI to fixed-size feature maps is by bilinear interpolation, and this enables utmost preservation of the feature maps' spatial information. **Figure 4** shows the framework of Mask R-CNN for instance segmentation.

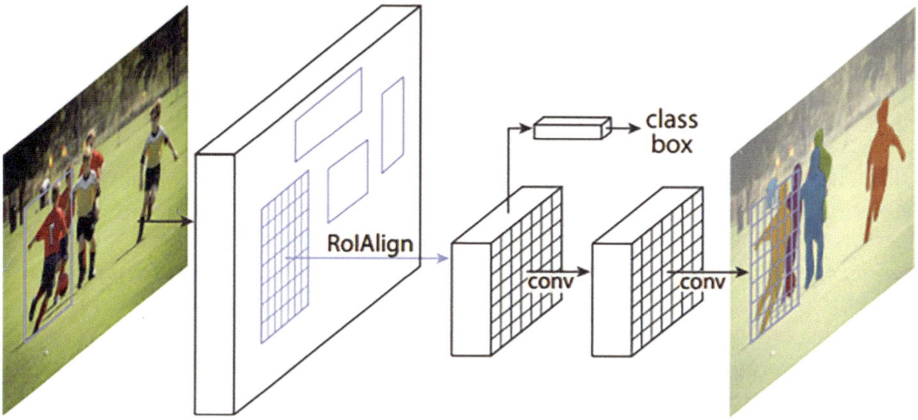

Figure 4.
The framework of Mask R-CNN for instance segmentation [11]. A two-stage procedure is adopted in the framework, the first stage comprises RPN and the second stage comprises a third branch for predicting segmentation masks on each RoI, in parallel with the classification branch and bounding box regression branch. The output of the additional mask is different from the outputs of the class and bounding box.

The RoI features alignment issue overlooked in RoIPool (RoI pooling) layer of Fast R-CNN and RoIWarp of MNC [46] was addressed by Mask R-CNN using RoIAlign layer, which eliminates the harsh quantization of RoIPool, accurately aligning the extracted features for the preservation of explicit per-pixel spatial correspondence. Cascade R-CNN [47] is a multistage extension of R-CNN, developed to improve object localization ability of the preceding models, and it improves accuracy at high Intersection over Union (IoU), addressing IoU mismatch. In most cases, to determine the positive and negative samples, the IoU threshold of the existing related works is set to 0.5, causing model to overfit. However, the bounding boxes of candidate objects are progressively refined at each stage of Cascade R-CNN using increasingly firmer IoU thresholds, permitting more object detection accuracy with tighter bounding boxes, especially at higher levels of IoU.

Basically, the first stage is used for rough object localizations, while the successive stages increasingly refine the bounding boxes with higher precision. Other notable examples of R-CNN series include Multi-Region CNN (MR-CNN) [48], Multi-Scale CNN (MS-CNN) [49], Adversarial Fast R-CNN (A-Fast-RCNN) [50], Libra R-CNN [51], Sparse R-CNN [52] and ME R-CNN [53].

2.2.2 YOLO series anchor-based object detection algorithms

The YOLO series is unarguably the most prominent and widely accepted family of object detection models among the CV community. Famous for striking a better balance between accuracy inference speed, the YOLO series demonstrates higher efficiency of one-stage object detection algorithms than the two-stage detection algorithms. It comprises both two-stage (anchor-based) and one-stage (anchor-free) approaches, indicating the series as a complete or near-complete family of object detection models. Chronologically, the YOLO series has approximately 11 core versions within the contexts that have shaped its sequential advancement. YOLOv1 [12] algorithm, which happens to be the first of its kind in YOLO series, was developed without an anchor box mechanism.

As an alternative, the input image was divided into S × S grid cells, and the class probabilities and bounding boxes were output on this grid cell. The extraction of features, the combination of those features, the classification and prediction of bounding box regression are all performed in the same CNNs. This unified architecture extremely improves the processing speed and optimization of YOLOv1. However, YOLOv1 has some notable limitations that include enforcement of spatial constraints on bounding box predictions because only one class prediction and two bounding boxes prediction are possible by this version of YOLO, making it extremely difficult for object generalization in unfamiliar configurations. Moreover, overlapping is an issue with YOLOv1, and it also struggles with object localization and prediction of multi-scale objects. **Figure 5** shows the detection system of YOLO.

YOLOv2 [54] improves the YOLOv1 model by adopting the anchor boxes using Batch Normalization (BN) on the input of each network's layer to enable the model's fast convergence. Five anchor boxes were positioned by YOLOv2 with different aspect ratios at the same location with the support of k-means clustering, to meliorate the detection model's recall rate for small objects. In this YOLOv2, the FC layer was replaced with FCN in the prediction stage for classification and bounding box regression. These developments improve the detection performance of YOLOv2 model with the assumption of maintaining high-processing speed. The YOLOv3 [55] model added several refinements to YOLOv2 for optimal performance of the network when extracting features.

In YOLOv3, an objectness score is predicted for each bounding box with the aid of logistic regression. Object's class is predicted by means of multi-label classification with individual logistic classifier as a substitute for softmax, helping in convoluted domains with several overlapping labels. DarkNet-53 was used as the backbone network of YOLOv3, and the backbone network comprises residual modules for efficient stacking of the deep network structure. Moreover, nine anchor boxes of three scales were positioned in the same position. Multi-scale object detection and multi-label classification problems are managed by replacing the activation function with sigmoid from softmax. There is a great improvement in detection accuracy and speed of YOLOv3.

The YOLOv4 [56] model incorporated several improvements over YOLOv3 model for object detection in real-time. DarkNet-53 of YOLOv3 was extended by the YOLOv4 backbone CSPDarkNet-53 with Cross-Stage Partial Connections (CSPNet) [57] structure for accurate training of the model for generalization ability and reduction of feature maps redundancy. CutMix [58], and Mosaic and Self-Adversarial Training (SAT) are used for data augmentation. Moreover, an improved Path Aggregation Network (PANet) [59], along with Convolutional Block Attention

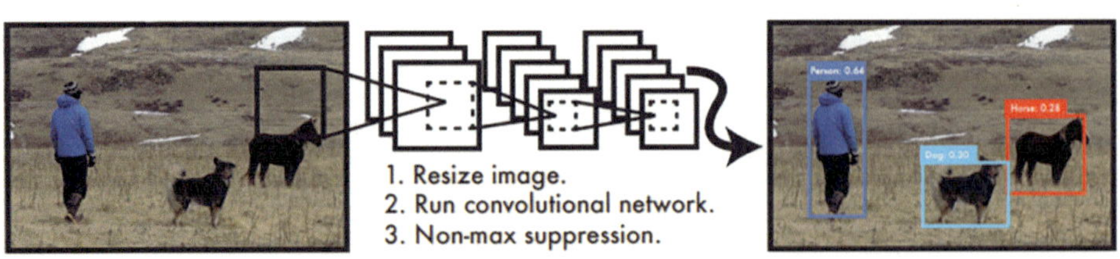

Figure 5.
The detection system of YOLO illustrates the simple and straightforward steps in processing images. (1) The input image is resized to 448 × 448 by the system, (2) a single convolutional network is run on the image and (3) the resulting detections are threshold by the model's confidence.

Module (CBAM), and SPP module are used by YOLOv4 on the neck part of the network for the improvement of spatial localization accuracy, achieving higher AP than YOLOv3.

However, YOLOv4 has a notable limitation, which is relatively large number of parameters. YOLOv5 [60] is another enhanced version of the YOLOv3 model, with basic similarity to the YOLOv4 detection network. The YOLOv5 model adapts CSPDark-Net53 as backbone. YOLOv5 uses different techniques of data augmentation such as MixUp [61], Copy-Paste [62] and other augmentations from Albumentations [63] collection. In the initial training stage of the YOLOv5 model, an adaptive anchor box screening and image scaling mechanism are proposed to speed up detection. YOLOv6 [64], an anchor-free object detection model, adopts an EfficientRep, motivated by RepVGG [65], for the backbone; and Rep-PAN (an enhanced PAN with RepBlock) with CSPStackRep Block (an enhanced CSP [66] block), for the neck; and a decoupled head using a hybrid-channel strategy, for the head.

While Task Alignment Learning (TAL) is employed for label assignments, the loss functions employed VariFocal Loss (VFL) for classification, and GIoU loss [67] or SIoU loss [68] is employed for bounding box regression. A more lightweight anchor-based YOLOv7 [69] model was proposed for the improvement of network efficiency. The model adopts Extended-Efficient Layer Aggregation Network (E-ELAN) and RepConvN, a variant of RepConv [70] for efficient learning of features. By this extension, the channel was expanded and computational blocks regularized *via* group convolution, without affecting how the parameter is utilized. In the same YOLOv7 model, two new label assignment approaches were proposed that are based on deep supervision [71] and trainable bag-of-freebies (BoF) methods, which include implicit knowledge and prearranged re-parameterized model, reducing the parameters of the model and improving the detection accuracy. The model also focuses on module optimization during the model training.

YOLOv8 [72] is an anchor-free object detection model, adopting the center-based pattern. Its backbone is like the backbone of YOLOv5, though with some modifications, which includes C2f, a variant of C2 module [73], which enables simple computation by processing the element that comes last on the split list by means of multi-bottleneck layers and results concatenation. The classification (made possible by Binary Cross Entropy (BCE)) and bounding box regression (made possible by CIoU loss [74] and DFL loss [75]) are independently handled by YOLOv8 through adoption of a decoupled head without including any objectness branch. A new framework, called Programmable Gradient Information (PGI) that assists in network supervision, was introduced in YOLOv9 [76] as a solution to input data volatility in deep networks, using its three components: the main branch for inference, the auxiliary reversible branch and multi-level auxiliary information.

Furthermore, a deep supervision fit for shallow and lightweight neural networks was presented by YOLOv9 model. The YOLOv9 model is an extension of YOLOv7 and its variant Dy-YOLOv7 [77], respectively. Generalized-ELAN (GELAN) replaces ELAN [78] in the network architecture design, simplifying the down-sampling module and optimizing the anchor-free prediction head. YOLOv10 [79] presents several architectural modifications, and its training is based on Non-Maximum Suppression (NMS)-free approach with multi-label assignments and matching metric consistency. An efficient partial self-attention (PSA) module design is implemented in the YOLOv10 model to integrate self-attention [80] excluding high computational complexity and traceable memory. A lightweight architecture comprising two depth-wise standalone convolutions was adopted by the classification head [81].

An efficient down-sampling is made possible by decoupling the spatial reduction and channel increase operations. Redundant stages complexity is reduced with the introduction of a rank-guided block design scheme. Other YOLO series include YOLOv11 [82], released by Ultralytics, with no specific papers written on its documentation. The PP-YOLO [83], an anchor-based object detection model, and its PP-YOLO series were named after PaddlePaddle [84], the platform on which they were developed. More PP-YOLO series include PP-YOLOv2 [85], which is an anchor-based object detection model too, and PP-YOLOE [86] and PP-YOLOE-R [87], both of which are based on anchor-free architecture. YOLOR [88] is an anchor-based object detection model that uses multi-task learning approach.

YOLOX [89] is another model that is based on anchor-free architecture and proposed immediately after YOLOR was released, with several improvements. DAMO-YOLO [90] was proposed as an anchor-free object detection model with several improvements integrated into it. YOLOX-PAI [91], an improved version of YOLOX, was presented to the CV community a year after DAMO-YOLO was released. Other variants of YOLO model are Gold-YOLO [92], YOLO-MS [93], YOLOCS [94] and YOLO-based Transformer, such as ViT-YOLO [95], MSFT-YOLO [96], NRT-YOLO [97], YOLO-SD [98] and DEYO [99].

2.2.3 SSD series anchor-based object detection algorithms

SSD series is one of the most influential families among one-stage object detection models since the development of SSD [100]. Like the YOLO series, SSD is a one-stage object detection model of dense prediction that is based on a single feed-forward convolutional network, which directly generates bounding boxes regression and confidence scores for the object class instances that are present in those bounding boxes, in accordance with the chosen anchor box of positive samples. **Figure 6** shows the framework of the SSD model. The major problem addressed by SSD is the small object detection problem confronting early YOLOv1 model. It achieves this by using convolutional layers of different sizes at the backbone network for multi-scale features extraction, excluding the FC layer in the prediction stage, while utilizing the FCN for the model parameter reduction.

SSD has the capacity for object detection at multi-scale across different layers of the network, unlike earlier object detection models that object could only be detected

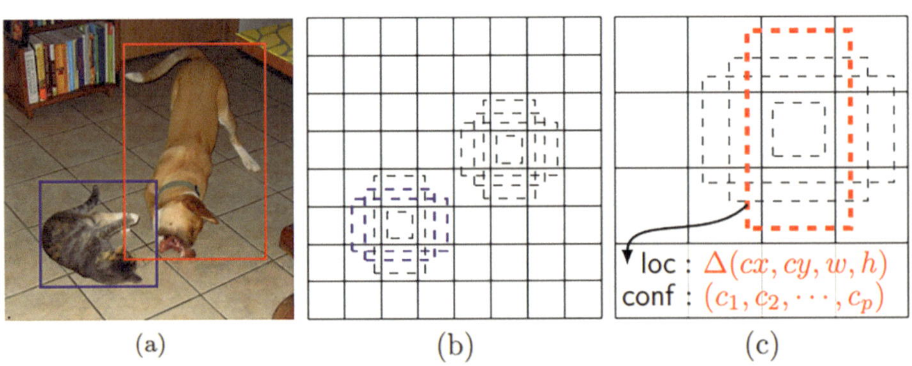

Figure 6.
Framework of SSD [100]. (a) Only input image and ground-truth boxes of each object are needed by SSD during training, (b) and (c) illustrate the different scales (e.g., 8 × 8 (b) and 4 × 4 (c)) during the training.

at the top layers of the network. However, there are some notable problems with SSD, among which are difficulty in detecting small objects due to unused context information among feature maps of different layers (this problem was subsequently solved by some models of SSD series [101–106], modifying the feature fusion module), and class imbalance among the samples during training due to shortage distribution of the generated anchor boxes near the real box.

Deconvolutional SSD (DSSD) [101] is an extension of SSD, designed with additional context for small object detection. It improves the model accuracy using ResNet-101 as the backbone network and builds the hourglass network architecture by adding a deconvolution module after the auxiliary convolutional layer, capable of obtaining a high-level, semantic- and low-level spatial information. Although DSSD demonstrates substantial and improved accuracy over SSD in detecting small objects, this comes with a reduced speed. Rainbow SSD (R-SSD) [102] improves the relationship between feature maps of different layers by employing pooling module and deconvolution module. The speed of DSSD and R-SSD models is slower than the speed of SSD model due to the many parameters involved in their network architectures. Extension modules were added to the first three convolutional layers by the ESSD [103] model for effectual possession of semantic features.

Feature-Fusion SSD (FSSD) [104] is another extension of SSD that integrates an efficient and lightweight feature fusion module. The primary reason for developing FSSD is to generate a feature pyramid, which is utilized for generating object detection outputs in SSD, by fusing all the different level features at the same time. The superiority of FSSD over SSD is demonstrated in its effectiveness for small object detection, which is very difficult for SSD to detect because it relies on shallow layers that have small receptive field for observing substantial context details. However, the inference time of FSSD is slower compared to SSD's inference time, due to additional layers added to the SSD model by FSSD, which consumes extra time.

Feature Fusion and Enhancement for SSD (FFE-SSD) [105], like FSSD, contributes to SSD enhancement by presenting multi-level feature fusion for the improvement of small object detection. Attentive SSD (ASSD) [106] was proposed as an extension of SSD with the idea of developing an attention model that can take notice of the connections among the pixel-level features. This was made possible by placing the self-attention mechanism between the feature map and the prediction module, which is the module designed for performing box regression and object classification. The Precise SSD (PSSD) [107] was proposed for the expansion of receptive field using dilated convolution. Moreover, real-time object detection was greatly improved with the spatial and semantic information obtained by the PSSD model using the bidirectional FPN architecture.

Another problem addressed by other models of SSD series is the notable problem associated with SSD, where the real box is provided with only a fractional portion of the created anchor boxes, leading to class imbalance between the data points that belong to the target class and the data points that do not belong to the target class during the model training process. To solve this problem, RetinaNet [108] was proposed. RetinaNet is a landmark among one-stage object detection models for its advanced approach to solving class imbalance, a notable problem with one-stage object detection models in achieving performance accuracy. In developing RetinaNet, a new loss function, called Focal Loss, was proposed, which has the capability of adding to the weight of hard samples in the cross-entropy loss during the model training process, maintaining class balance between the data points that belong to the target class and the data points that do not belong to the target class.

RetinaNet was enhanced by RetinaMask [109] with several adjustments, which include a self-adjusting Smooth L1 loss, instance mask prediction integration and the extra hard examples addition during model training process. RetinaNet was enhanced by Retina U-Net [110] with notable improvements like incorporation of U-Net [111] for segmentation of biomedical images. A rescaled member of RetinaNet was proposed by RetinaNet-RS [112], increasing the speed of the model but decreasing in its accuracy. RefineDet [113], a sparse prediction model, was proposed to effectively address the problem of class imbalance by applying anchor box refinement module for removing negative samples and fixing the location and magnitude of positive samples. However, there is a problem with RefineDet-derived box multi-scale features aligning and optimizing the anchor box properly.

To address this problem, Enriched Feature Guided Refinement Network (EFGRNet) [114] model was proposed. EFGRNet enhances SSD by introducing feature enrichment (FE) scheme and a cascaded refinement scheme for solving the multi-scale detection and class imbalance problems, obtaining accurate prediction results without affecting its speed.

2.3 Anchor-free object detection algorithms

Although the anchor-based object detection algorithms allow positive sample selection for box regression and final classification in object detection process, relying on predefined boxes with fixed sizes and aspect ratios, they have some notable limitations like class imbalance between the data points that belong to the target class and the data points that do not belong to the target class due to more negative samples of dense anchor boxes generated. Also, there are problems of object's localization and background complexity with the anchor box of the data points that belong to the target class.

Furthermore, several hyperparameters were introduced by anchor boxes, such as the quantity of anchors, the sizes and aspect ratios, which require extemporary rules-of-thumb and computations from training set for their tuning, and the tuning becomes a herculean task when integrated with multi-scale architectures. To address these problems, anchor-free object detection algorithms have been proposed, eliminating pre-defined anchor boxes. Anchor-free object detection models can be classified into keypoint-based and anchor-point-based (otherwise known as center-based) approaches.

2.3.1 Key-point-based anchor-free object detection models

Keypoint-based anchor-free object detection models detect image objects using multi-predefined keypoints or self-learned keypoints like the image center, corners and sparse key corners, which by grouping them enables bounding box prediction, as shown in **Figure** 7.

The first model in this category is CornerNet [116], which uses pairs of keypoints in terms of bounding box's top-left and bottom-right corners to detect objects. In addition, the pooling layer was introduced by CornerNet to enable better localizations of bounding box corners. In the following year, the efficiency of CornerNet's inference was improved by CornerNet-Lite [117] with the introduction of CornerNet-Saccade, which introduces the attention mechanism for reducing number of pixels processed, and CornerNet-Squeeze, which reduces the pixel computation.

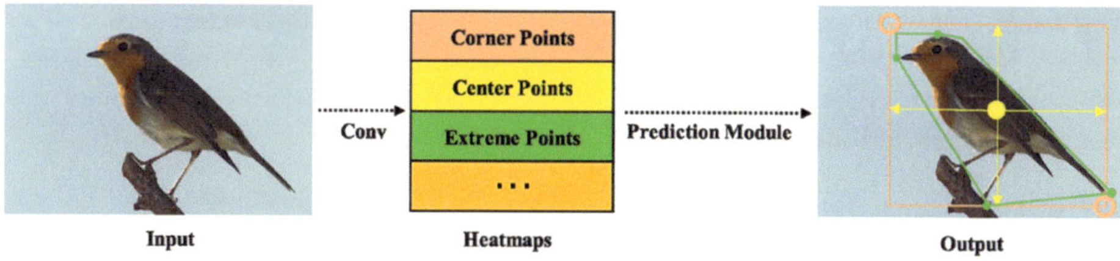

Figure 7.
Key point-based anchor-free object detection model [115].

However, CornerNet has some limitations such as loss of feature information to some degree due to the corner pooling layer relationship with the max pooling process. This motivated the development of Matrix Nets (KP-xNets) [118] and deep neural network for object detection. Objects with different sizes and aspect ratios are mapped by KP-xNets into layers ensuring uniformity or near uniformity of the objects' sizes and their aspect ratios within their layers. Therefore, KP-xNets offer a scale and aspect ratio informed architecture. KP-xNets were leveraged for the enhancement of key-points-based object detection. The network architecture of KP-xNet is lightweight, with an improved detection speed. CenterNet [119] resolves the limitation in CenterNet by adding a third keypoint to the existing two keypoints at the bounding box's geometric center, making them three, instead of two, for objects' representation.

The added point improves the model's capability for capturing the visible patterns in the proposed regions and confirming the accuracy of each bounding box. Hence, CenterNet advances CornerNet by employing an alternative representation for prediction box, through which bounding boxes are reproduced by generating for the object, a heatmap of the center point in addition to the corresponding size. With CenterNet, it is easy to remove the superfluous prediction boxes on the heatmap without wasting time, unlike the NMS post-processing operation. Moreover, two pooling approaches, center pooling and cascade corner pooling, were proposed for the improvement of keypoints detection of image's center and corners.

ExtremeNet [120], an anchor-free box detection model, was proposed as a solution to many immaterial background pixels that may contain in the prediction box, hindering true representation of the object's visual information. ExtremeNet uses hourglass network for the detection of all the object's four heatmaps (i.e., top, bottom, left and right four extreme points) and picks out the right combination of five extreme points in accordance with the four extreme points' geometric center's scores on the central point heatmap. That is, the extreme points were extracted as peaks after the five heatmaps were predicted and geometrically grouped. The geometric center of the four extreme points is computed, and the extreme points are regarded as the true detection provided there is a high prediction response in the center of the center map.

Other key-point-based object detection models are AttentionNet [121] that continuously improves bounding boxes through directional shifts prediction for the object's corner keypoints; Point Linking Network (PLN) [122], which uses corners, center points and their links to represent objects; and CentripetalNet [123] introduces centripetal shift method (a method for grouping corner keypoints from same instance) for corner matching improvement. DeNet [124] is a keypoint-based anchor-free object detection model that produces RoIs without anchor boxes. DeNet, in an identifiably distinctive manner, initially ascertains the likelihood of each position being a member of any of the bounding box's several corners and subsequently produces RoIs by calculating all possible corner combinations, adopting two-stage approach for the classification of each RoI. Other popular keypoint-based object detection models include Grid R-CNN [125] and RPDet [126].

2.3.2 Anchor-point-based anchor-free object detection models

Anchor-point-based object detection models use preset anchor points to achieve object detection in a per-pixel prediction fashion. DenseBox [127] is a foremost figure among anchor-point-based anchor-free object detection frameworks, introduced purposely for face detection. In DenseBox framework, output ground-truth is defined as a 5-channel map, and it locates objects by predicting the distance between each positive pixel and boundaries of the bounding boxes. Afterward, several anchor-point-based object detection models have been developed. The Fully Convolutional One-Stage (FCOS) [128] model was proposed as a representative model, adopting multi-level prediction with FPN, and by using two branches, it enforces classification of both positive anchor and negative anchor points, and anchor point to prediction-box distance regression.

To subdue the anchor points interference with the center of the object on the regression, a centered branch for prediction is added to the classification branch. The FCOS model is both computationally fast and capable of being easily expanded or upgraded on demand to fit other vision problems. However, there are some constraints imposed by anchor boxes in FCOS model that hinder optimal feature-level selection for individual object instances. Feature Selective Anchor-Free (FSAF) [129] was designed to solve the constraints imposed by anchor boxes in FCOS model by selecting optimal feature level for individual object instances. To be specific, it is possible to plug the FSAF module into anchor-free object detection models with FPN, such as RetinaNet. In the end, the results are obtained by combining the prediction results from the two branches in the inference stage.

FCOS [130], an improved version of FCOS [128], was proposed for better refinement of the framework. BorderDet [131] extends FCOS by presenting BorderAlign, a feature extractor developed for capturing border features to improve the original key-point-based representation. CenterMask [132] was proposed as an extension of FCOS and added a spatial attention-guided mask (SAG-Mask) branch for segmentation of object instances. FoveaBox [133] was proposed to solve the limitations of the earlier anchor-point-based object detection models, in which each instance occupies separate feature layer. In FoveaBox, the acceptable range of sizes for each layer of FPN is predetermined. The instance is assigned to the appropriate multi-feature layers for learning in accordance with the ground-truth size.

The prediction box is output by regressing the distance between the pixel point and corner keypoints. The architecture of anchor-point-based was designed as a lightweight and efficient network, making it better than the key-point-based method. However, it has a lower detection accuracy. The Soft Anchor-Point Detector (SAPD) [134] model adopts a training strategy with soft-weighted anchor points and soft-selected pyramid levels to control anchor points close to instance border from partaking in the training process, achieving an equilibrium between efficiency and effectiveness. The prediction boxes of the key-point-based object detection models and anchor-point-based object detection models are created by discrete points.

There are some notable challenges with prediction boxes such as inability to keep adjacent information from loss, and difficulty in creating the required extra branches for the pairing of discrete points from the same object. CrossDet [135] was proposed as a solution to SAPD challenges by adopting two adaptive crossing lines for the object's position representation. The rough location of the cross lines is predicted by CrossDet through a regression branch before aggregating the line features, in a way that involves changing the features to suit the changing conditions, from the dimensional crossing lines through a crossline extraction module, and the location of the crossing lines is regressed through a decoupling technique. Lightweight object detection models, such as PP-PicoDet [136], on mobile phones were motivated by FCOS for improved real-time object detection. **Figure 8** shows the steps involved in anchor-point-based anchor-free object detection model.

2.3.3 YOLO series anchor-free object detection algorithms

The YOLO series object detection algorithms were improved for anchor removal using key-point-based and anchor-point-based object detection methods. YOLOX [89] is another model that is based on anchor-free architecture and proposed immediately after YOLOR [88] was released, with several improvements. YOLOv3 was used as the baseline for YOLOX, with the input image split into multi-grids, whereby the prediction of the two offsets in the corners and entire size of the prediction box can be done at once grid-by-grid. Using YOLOX, end-to-end network parameters are easily optimized by separating regression tasks from classification. Another spectacular contribution of YOLOX is the use of SimOTA, which necessitates label assignment by ground-truth objects during training and not during inference time.

OTA is a method that handles global context label assignment and assists in finding the global best confidence assignment for all image instances. YOLOX-PAI [91], an improved version of YOLOX, was presented to the CV community a year after DAMO-YOLO [90] was released. In the head, YOLOX-PAI uses attention mechanism for organization of tasks in the regression and classification branches and uses an adaptive spatial feature fusion strategy for features enhancement. PP-YOLOE [86], which is based on anchor-free architecture, carries out pixel-level prediction using predetermined anchor points, which significantly regulates the model parameters and improves the inference speed. Moreover, by introducing TAL, PP-YOLOE attempts to close the distance existing between the optimal anchor points of classification and regression tasks, thereby getting prediction results with high localization and confidence accuracy, simultaneously.

PP-YOLOE-R [87], which is also based on anchor-free architecture, was proposed in the same year as PP-YOLOE [86] as an improved model for the detection of rotating

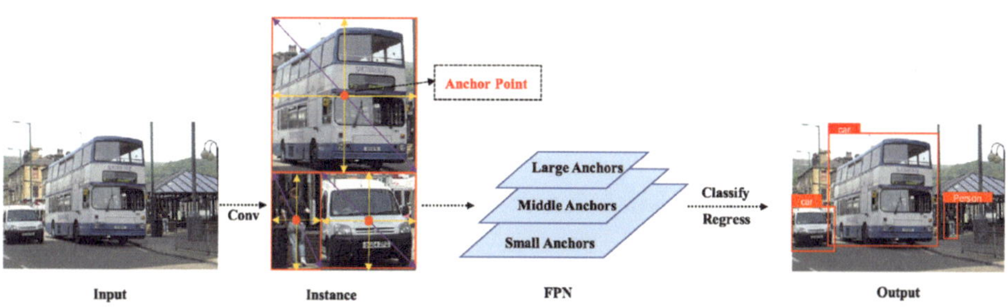

Figure 8.
Anchor-point-based anchor-free object detection model [115].

objects. The disconnection problem in object boundary is handled by the ProbIoU loss and it uses decoupling detection head for the realization of object angle's prediction. DAMO-YOLO [90] was proposed as an anchor-free object detection model with several improvements integrated into it. Other variants of YOLO model are Gold-YOLO [92], YOLO-MS [93], YOLOCS [94], and YOLO-based Transformer, such as ViT-YOLO [95], MSFT-YOLO [96], NRT-YOLO [97], YOLO-SD [98], and DEYO [99].

Several other one-stage object detection models have been proposed with high-performance accuracy. G-CNN [137] presents a grid-based object detection model that uses a series of iterations in refining a static multi-scale grid of bounding boxes for objects' localization and classification. Reverse connection with Objectness prior Networks (RON) [138] uses reverse connection to improve the location prediction of multi-scale objects and presents the objectness before reducing the search space for objects. While Scale-Transferrable Detection Network (STDN) [139] uses high-resolution and scale-transfer layers, TripleNet [140] employs an encoder-decoder approach for the detection and semantic segmentation of fused objects.

Dual Refinement Network (DRNet) [141] integrates an anchor-offset technique with feature localization and anchor refinement and a warp detection head. EfficientDet [142] introduces a bi-directional feature pyramid network (BiFPN) for the fusion of multi-scale features. It also proposes a method (compound scaling) for scaling the neural network's resolution, depth and width. The method enables EfficientNet architecture to generate models that are accurate and efficiently utilize computational resources. Reciprocal Object Detection and Instance Segmentation Network (RDSNet) [143] improves RetinaNet with two CNNs for object detection and instance segmentation. While spatial features are processed by one CNN, temporal features are processed by the other.

Task-aligned One-stage Object Detection (TOOD) [144] employs Task-aligned Head (T-Head) and TAL for alignment of classification and localization tasks. DetectorNet [145] and OverFeat [146] are among the first generation of several other one-stage object detection models. DetectorNet adopts AlexNet as backbone and divides the input image into a grid (coarse) and structures the detection (regression) problem to bounding box masks, while Overfeat represents a merged framework for detection, localization and classification.

2.4 Transformer-based object detection algorithms

Features are extracted by CNNs in neural network operation by sliding individual convolutional windows. This operation somewhat incapacitates the model's from obtaining global feature information. The above challenges are effectively addressed by the Transformer-based object detection algorithms. Global context can be modeled effectively by the Transformer algorithms using self-attention mechanism, a deep learning technique that models use to understand the relationships between words in a sequence. The Transformer-based object detection algorithms have two main series, DETR series and ViT series.

2.4.1 DETR series transformer-based object detection algorithms

DETR [15] is the first among the Transformer-based object detection algorithms and upon which DETR-series were built. Convolution module is utilized by DETR for back bone features extraction, and it uses neck situated Transformer structure for image features encoding into a set of learned positional embeddings as shown in **Figure 9**. By set decoding and prediction, each object's category and location information are obtained. The global context relationship between the object and the image is better managed by DETR, and for each object, it uses Hungarian algorithm [147] for single-bounding box prediction, removing the unnecessary steps involved in applying NMS for filtering out redundant bounding boxes that have high overlap, which is a common limitation of modern object detection models.

However, DETR has some notable drawbacks. Comparing DETR to modern object detection models, a substantial amount of training epochs is required by DETR to reach convergence, and this is primarily due to the difficulty involved in training the attention modules. In addition, the detection performance of DETR on small objects is rather low and this is due to its inability to rely on high-resolution feature maps, caused by self-attention module's quadratic complexity in the Transformer encoder, thereby making high-resolution input images computationally unaffordable.

In **Figure 9**, the image is flattened and supplemented with a positional encoding by the model before it is passed into the encoder of the transformer. A small, fixed number of learned positional embeddings is then taken as input by the decoder of the transformer, and it also takes care of the encoder output. A shared feed forward network (FFN) handles the decoder's each output embedding that is passed to it for prediction of either a detection of object class (including the bounding box) or a no object class. To solve these problems, some methods for improvement have been proposed including attention module and queries improvement. Adaptive Clustering Transformer (ACT) [148] employs an effective randomized technique (locality sensitive hashing) for grouping similar queries features for data clustering.

In DETR, it is possible to reduce floating point operations per second (FLOPs) without losing performance by substituting the self-attention module with ACT. SMCA [149] module was proposed as a replacement to the co-attention module in DETR and it minimizes the feature aggregation's search range to close to the object's center. Moreover, SMCA speeds up the convergence by joining the learnable co-attention weights with the query's spatial prior. The limitations of DETR such as the resolution of small features and the delay in training convergence (which is caused by the variability in matching discrete bipartite graphs) were overcome by the dynamic attentions introduced by dynamic DETR [150] in the encoder and decoder stages of DETR. Deformable DETR [151] was introduced as a multiple-scale deformable attention model to overcome DETR's limitations.

Deformable DETR design improves the detection of small objects by combining the deformable convolution's spatial sampling capability with Transformer's global relational modeling capability for performance improvement of small object detection. The speed at which Deformable DETR converges is also improved due to a few

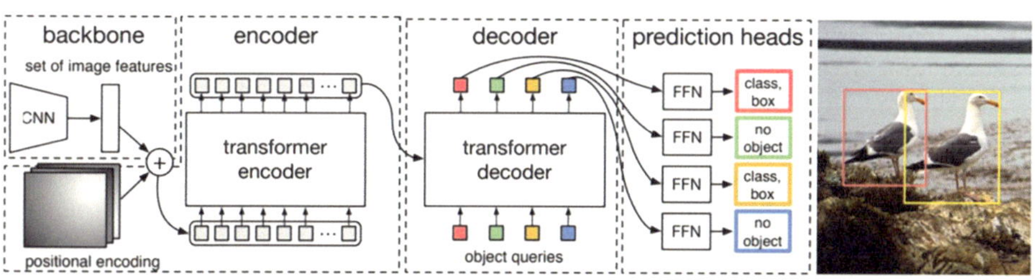

Figure 9.
The framework of DETR [15]. DETR employs CNN's backbone for learning input image's 2D representation.

sampling keypoints needed by the model. However, the Deformable DETR model becomes more complex, resulting from the model feeding the encoder with multi-scale features indiscriminately. Sparse DETR [152] was proposed to minimize computational constriction in DETR and Deformable DETR by introducing a mechanism for encoder token sparsification and auxiliary loss function. Sparse DETR filters the encoder's tokens (input units) using a scoring network and converts the features into code with the selected superior tokens, thereby obtaining sparser queries and improving inference speed.

Dynamic Anchor Box DETR (DAB-DETR) [153] contributes to the decoder of DETR and speeds up the convergence of the model by adopting dynamic anchor boxes (box coordinates) in the Transformer decoders, which are dynamically updated layer by layer as queries for position priors provision. Motivated by Conditional DETR [154], DAB-DETR combines location and contents to form queries, with the cross-attention module keyed in, separating the content and location inputs to the query-to-feature resemblance. For the location embeddings to be re-scaled, conditional spatial query is carried out. 2D coordinates are utilized in generating the location embeddings in queries and keys, making their update possible layer by layer.

Anchor DETR [155] presents a new approach of deriving object queries using anchor points.

The queries are encoded by Anchor DETR by predetermining the anchor points and surrounding them with multi-prediction branches, and it figures out the object's overlapping occlusion problem. Moreover, a Row-Column Decouple Attention (RCDA) module is presented by Anchor DETR to uncouple the 2D feature map into row-column features for cost-effective computation. DeNoising DETR (DN-DETR) [156], built upon DAB-DETR, was proposed as an advanced method for training model using a query denoising task for stabilizing the matching of bipartite graphs during training.

DETR with Improved deNoising anchOr boxes (DINO) [157] combines the characteristics of DN-DETR, DAB-DETR and Deformable DETR, for DETR's improvement by providing the model with improved denoising anchor boxes, thereby assisting the model in stabilizing the matching of bipartite graphs during training for overall computation efficiency. In furtherance of the contributions of DETR series Transformer-based object detection algorithms as presented earlier, some other contributions have been proposed in addition to the abovementioned ones to address the limitations of DETR model. Unsupervised Pre-train DETR (UP-DETR) [158] presents a random query patch detection for DETR pre-training. Efficient-DETR [159] presents efficient mechanism for recognizing arbitrary initialization of object containers and queries, and reference points, as part of causative factors for multiple iterations.

Rethinking Transformer-based Set Prediction DETR (RTSP-DETR) [160] presents Transformer-based Set Prediction with FCOS (TSP-FCOS) and Transformer-based Set Prediction with R-CNN (TSP-RCNN) as the efficient techniques for identifying the Hungarian loss and the cross-attention mechanism as the causative factor in DETR's slow convergence. Transformer-based detector Without Backbone (WB-DETR) [161] presents a DETR model without a CNN backbone, justifying the inconsequential contribution of feature extraction based on CNN in transformer-based object detection models. Poll and Pool DETR (PnP-DETR) [162] adopts a poll and pool (PnP) sampling module for spatial redundancy reduction, thereby improving the efficiency of DETR model. Decoder-Only DETR (D^2ETR) [163] eliminates the encoder to make DETR simpler and presents a Computationally Efficient Cross-scale Attention (CECA) module.

Fully Pre-training DETR (FP-DETR) [164] was proposed as a method for complete pre-training an encoder-based transformer and fine-tuning it for object detection using a task adapter. Inspired by the realization of textual prompts in Natural Language Processing (NLP), query positional embeddings were treated as visual prompts so that the model can take care of the target area and identify the object. DETR was improved by Coarse-to-Fine DETR (CF-DETR) [165] by using coarse-to-fine (CF) decoder layer for coarse features refinement and location prediction. Particularly, the layer of the CF decoder comprised a coarse layer and a cautiously created fine layer. In each layer of the CF decoder, the region of interest feature is presented to the global context information flow from the coarse layer for the refinement and enrichment of features of the object query through the fine layer.

Recurrent Glimpse-based decOder (REGO) [166] uses a multi-stage recurrent computational construction to assist the DETR attention slowly concentrate on objects in the foreground more accurately. In all the processing stages, visual features were extracted from RoIs as glimpse features with expanded bounding box detection result areas from the earlier stage. Consequently, a decoder based on glimpse was introduced to help with the refined detection results based on both the glimpse features and the earlier stage's attention modeling outputs. Co-DETR [167] is a training scheme comprising collaborative hybrid assignments for learning DETR-based object detection models more efficiently and effectively from adaptable label assignment manners.

The learning ability of the encoder in end-to-end object detection models can easily be enhanced by the training scheme by training the multi-parallel heads (auxiliary) superintended by one-to-many label assignments such as ATSS and Faster RCNN. MDETR [168] is an end-to-end modulated detector for image object detection trained on a raw text query such as a question or a caption. In MDETR, an architecture based on Transformer is employed for collaborative effort on text and image by combining the two modalities in the model's early stage. RefineBox [169] is a general framework conceptually made simple and efficient for localization challenges in DETR-based models. Instead of wasting time designing and training new models, plugins were added to the existing well-trained models. The outputs of DETR-based object detection models were refined by RefineBox using lightweight refinement networks. The implementation and training of RefineBox is simple because its features and predicted boxes are leveraged by those of well-trained detection models.

2.4.2 ViT series

Vision Transformers (ViT) [170] was proposed as a solution to problems confronting image classification and object detection tasks. The ViT models classify images by enlarging the range of pixel selection and splitting the input image into sparse patches. The ViT series represents a family of object detection models where ViT, primarily designed for image classification, is adapted to object detection tasks. **Figure 10** illustrates the framework of the ViT Model.

The backbone of ViT was extended by ViT-FRCNN [171] for the detection of objects by integrating a detection network based on Faster R-CNN framework. However, ViT-FRCNN has challenges using ViT as its backbone because the feature

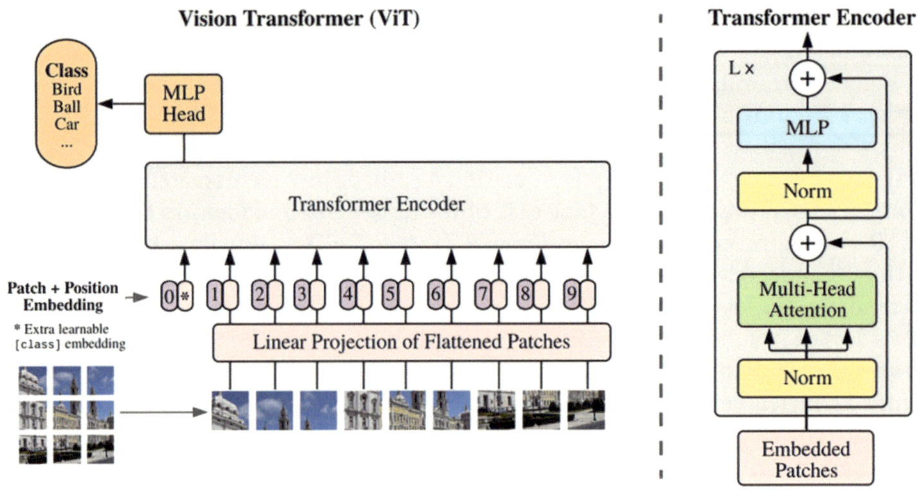

Figure 10.
ViT Model Framework [170]. An image was divided into fixed-size patches, with each of them linearly embedded, position embeddings added and feed the resulting vector sequence fed to a standard encoder of Transformer.

maps generated by ViT are not more than one scale and lack high resolution, thereby increasing the computational cost and image resolution. The training of ViT-FRCNN involves more high resolution than ViT, to keep enough resolution, and to maintain the input image's aspect ratio. Pyramid Vision Transformer (PVT) [172], a variant model of ViT, was proposed to strengthen a decreasing pyramid structure into the backbone of the Transformer for flexible learning of multi-scale high-resolution features. The spatial reduction attention (SRA) replaces the multi-head attention in ViT, making it computationally possible to process at low cost, the high-resolution feature maps and global receptive fields.

However, the images are regarded by PVT as non-overlapping patches sequence, affecting the feature's local continuity. The patch windows were expanded and overlapped by PVTv2 [173] to strengthen their connection and the feature map's resolution was maintained using zero-padding convolution. The computational complexity linear growth is monitored by Swin Transformer [18] by reducing the computation of self-attention to non-overlapping local windows, and slowly increasing the receptive field during the downsampling based on hierarchy, to facilitate the efficient feature extraction from local to global. Swin Transformer V2 [174] presents the methods for scaling Swin Transformer up to 3 billion parameters, enabling it to train with 1536x1,536 image resolution. With the Swin Transformer V2, issues of training instability were tackled, and models pre-trained were effectively transferred to higher resolutions from low resolutions.

ViTDet (ViT Detector) [175] adapts the backbone network of original ViT for object detection by fine-tuning its architecture without redesigning the backbone for pre-training. With this development, competitive results can be achieved by the plain-backbone ViTDet. Amazingly, it was observed that a sufficiently simple feature pyramid can be built from a feature map with single-scale without the FPN design, and window attention, assisted with exiguous cross-window propagation blocks, can be sufficiently employed. With plain ViT backbones such as ViT-B, ViT-L and ViT-H, pre-trained as Masked Autoencoders (MAE) [176], ViTDet is on a par with any hierarchical backbone-based methods. YOLOS [177] is a family of vanilla ViT-based object detection models, with minimal modifications to enable Transformer carry out 2D object-level and region-level recognitions from a clean Seq2Seq (sequence-to-sequence) view with minimal cognition of the spatial structure.

Multiscale Vision Transformers (MViT) [178] presents a computationally cheap method for recognizing videos and images, by incorporating feature hierarchies (multiscale) with transformer models. The temporal dimension was removed for successful application of MViT to image classification. MViTv2 (Improved MViT) [179] was presented as a combined architecture for object detection, and video and image classification. MViTv2 extends MViT by incorporating disintegrated relative positional embeddings and residual pooling mechanism. XCiT (Cross-Covariance Image Transformers) [180] is built upon XCA (cross-covariance attention) and is a permutation-like version of self-attention that spreads its operation across feature channels rather than the Transformer's tokens. The output XCA possesses linear complexity in the token's number and permits efficient computation of high-resolution images. The conventional transformer accuracy is combined with the convolutional architectures' scalability. **Table 1** presents the summary of the evolution of 2D object detection algorithms from the AlexNet era to Transformer-based era.

Object detection algorithms		**Author(s) & years**
Anchor-based	R-CNN	Girshick et al. [9]
	SPPNet	He et al. [34]
	Fast R-CNN	Girshick et al. [36]
	Faster R-CNN	Ren et al. [10]
	R-FCN	Dai et al. [21]
	YOLOv1	Redmon et al. [12]
	SSD	Liu et al. [13]
	Mask R-CNN	He et al. [11]
	FPN	Lin et al. [19]
	YOLOv2	Redmon and Farhadi [54]
	DSSD	Fu et al. [101]
	R-SSD	Jeong et al. [102]
	YOLOv3	Redmon and Farhadi [55]
	Cascade R-CNN	Cai and Vasconcelos [47]
	RefineDet	Zhang et al. [113]
	ESSD	Zheng et al. [103]
	FSSD	Cao et al. [104]
	ASSD	Yi et al. [106]
	YOLOv4	Bochkovskiy et al. [56]
	YOLOv5	Jocher et al. [60]
	PSSD	Chandio et al. [107]
	YOLOv7	Wang et al. [69]
	YOLOv12	Tian et al. [181]

Object detection algorithms		**Author(s) & years**
Anchor-free	CornerNet	Law and Deng [116]
	FSAF	Zhu et al. [129]
	CenterNet	Duan et al. [119]
	FCOS	Tian et al. [128]
	KP-xNet	Rashwan et al. [118]
	ExtremeNet	Zhou et al. [120]
	EFGRNet	Nie et al. [114]
	SAPD	Zhu et al. [134]
	YOLOX	Ge et al. [89]
	PP-YOLOE	Xu et al. [86]
	PP-YOLOE-R	Wang et al. [87]
	YOLOv6	Li et al. [64]
	YOLOv8	Jocher et al. [72]
	YOLOv9	Wang et al. [76]
	YOLOv10	Wang et al. [79]
	YOLOv11	Jocher and Qiu [82]
Transformer-based	DETR	Carion et al. [15]
	ACT	Zheng et al. [148]
	Deformable DETR	Zhu et al. [151]
	ViT-FRCNN	Beal et al. [171]
	Swin Transformer	Liu et al. [18]
	Sparse DETR	Roh et al. [152]
	PVT	Wang et al. [172]
	Swin Transformer 2	Liu et al. [174]
	PVTv2	Li et al. [173]
	DINO	Zhang et al. [157]
	DAB-DETR	Liu et al. [153]
	DN-DETR	Li et al. [156]
	ViTDet	Li et al. [175]

Table 1.
The evolution of 2D object detection algorithms.

3. Datasets

A brief overview of the competitively employed dataset baselines for generic 2D object detection are presented in this section. Since large datasets are required for evaluating the training and testing of object detection models, we also provide a summary of their main characteristics, including Pascal Visual Object Classes (Pascal VOC) [182], ImageNet [183], MS COCO [184], Open Images [185], Objects365 [186] and SA-1B (Segment Anything-1 Billion) [187]. The Pascal VOC challenge was an annual CV competition series, held between 2005 and 2012. VOC-2007 [188] and

VOC-2012 [189] are the two dataset versions of this challenge that are generally considered and used as standard benchmark for object detection. There are 5 k training images in VOC-2007 dataset, and over 12 k objects were labeled from 20 object classes.

Both the training and labeled images of VOC-2007 were increased to 11 k training images and 27 k labeled images in VOC-2012, respectively, without changing the number of their object classes. The ImageNet Large Scale Visual Recognition Challenge (ILSVRC) was also an annual competition series, held between 2010 and 2017. There are 460 k training images from 200 classes contained in the ILSVRC dataset, and all of them were selected from ImageNet [190]. The ILSVRC datasets

Figure 11.
Example of masked images from SA-1B dataset. The images were grouped for visualization based on the number of masks per image.

are larger in magnitude than the Pascal VOC datasets. MicroSoft Common Objects COntext (MS COCO) is one of the most standard large-scale and benchmark datasets for object detection and segmentation. With 80 object classes, MS-COCO has continued in popularity since its first release in 2014 with its modified version in 2017.

Open Images are Google's image dataset of approximately 9 M annotated images, covering the object bounding boxes and segmentation masks, etc. Publicly made available since 2016, Open Images have expanded into seven versions, with the last version comprising 1.9 M images of 600 object classes with 16 M bounding boxes. Objects365 was released in 2019 as a dataset for large-scale object detection. It comprises 600 k images of 365 object classes with over 10 M bounding boxes. SA-1B comprises over 1B masks from 11 M images (see **Figure 11**). SA-1B was collected by automatic technique using the final stage of Segment Anything Model (SAM) data engine. The masks are 400× more than any existing dataset for segmentation, with high resolution, quality and diversity.

The following analysis summarizes the abovementioned datasets for 2D object detection. Each version of the datasets has its number of classes, number of training images, number of validation images, and number of testing images. The parenthesized numbers denote the number of annotated instances. Pascal VOC-2007 has 20 classes, 2501 (6, 301) training images, 2510 (6, 307) validation images and 4, 952 testing images. Pascal VOC-2012 has 20 classes, 5717 (13,609) training images, 5823 (13,841) validation images and 10,991 testing images. ILSVRC-2014 has 200 classes, 456,567 (478,807) training images, 20,121 (55,502) validation images and 40,152 testing images. ILSVRC-2017 has 200 classes, 456,567 (478,807) training images, 20,121 (55,502) validation images and 65,500 testing images.

MS-COCO-2014 has 80 classes, 82,783 training images, 40,504 validation images and 40,775 testing images. MS-COCO-2017 has 80 classes, 118,287 training images, 5000 validation images and 40,670 testing images. OpenImages-v7 has 600 classes, 1,743,042 (14,610,229) training images, 41,620 (303,980) validation images and 125,436 (937,327) testing images. Objects365–2019 has 365 classes, 600,000 (9,623,000) training images, 38,000 (479,000) validation images and 100,000 (1,700,000) testing images.

4. Evaluation metrics

Several datasets with their corresponding 2D evaluation metrics have been launched in the history of CV and object detection for performance measurement of algorithms. Among these metrics are those for evaluating model accuracy such as Precision (P), Recall (R), Average Precision (AP), overall Average Precision (mAP) and Intersection over Union (IoU). Precision is used for measuring the percentage of objects correctly detected by the model (i.e., True Positives (TP)) among all the detected objects (Eq. (1)). Recall is used for measuring the percentage of objects correctly detected by the model relative to all the positive objects (Eq. (2)). AP is employed for accuracy evaluation of a class detector for a particular class calculation, which is calculated as the area under the Precision-Recall curve (Eq. (3)).

The mAP denotes the AP's average across all the classes (Eq. (4)). Both mAP and AP rely on the selected threshold of IoU. IoU is used for calculating the ratio of the predicted bounding box's intersection area (P_{bb}) to ground-truth bounding box's union area (G_{bb}) (Eq. (5)). A bounding box prediction is considered as positively predicted provided its IoU exceeds a predefined threshold with the ground-truth, or else, it is considered negative. To overcome the challenges of a standardized 0.5 IoU threshold, and the Pascal VOC's standardized mAP@0.5 (among which is lack of generality for multi-scale object detection), the MS-COCO dataset presented mAP@[0.5:0.95], denoting the mAP average for the thresholds of IoU between 0.5 and 0.95 at 0.05 intervals. Average Recall (AR) and Localization Recall Precision [191] are among other unpopular evaluation metrics. Efficient evaluation metrics, such as model inference time, parameters, Frame Per Second (FPS), are notable essential metrics for measuring the performance of real-time object detection models.

$$P = \frac{True\ positive}{True\ positive + False\ positive} \tag{1}$$

$$R = \frac{True\ positive}{True\ positive + False\ negative} \tag{2}$$

$$AP = \sum_{n=1}^{N} \left[R(n) - R(n-1)\right].maxP(n) \tag{3}$$

$$mAP = \frac{1}{n}\sum_{n=1}^{N} AP_i \tag{4}$$

$$IOU = \frac{\mathrm{P}_{bb} \cap \mathrm{G}_{bb}}{\mathrm{P}_{bb}\ \mathrm{U}\ \mathrm{G}_{bb}} \tag{5}$$

5. Future research directions

As technology advances, there are many challenges confronting the implementation and deployment of object detection models. Some of the optimistic research directions are highlighted in this section, expected to address these challenges and smoothen the future of object detection.

5.1 Neural architecture search

The frameworks of conventional detection networks are mostly manually designed, which often lead to a tradeoff between accuracy and efficiency. By formulating an automated iteration-driven search approach in a stated search space, optimistically, this problem can be addressed by progressive neural architecture search (NAS) [192], to achieve optimal network model accuracy and efficiency. Therefore, a promising direction for future research is by employing NAS for automated design or detection network optimization. Among the several object detection networks that have applied NAS are DetNAS [193], One-shot Path Aggregation Network Architecture Search (OPANAS) [194], NAS-FPN [195], Hit-detector [196], PP-PicoDet [136], Auto-FPN [197], SM-NAS [198], NAS-FCOS [199], SpineNet [200], FBNetV5 [201], MobileDets [202]. As models for object detection keep expanding in scale and NAS keeps progressing, it is expected that NAS will circumvent the manually designed conventional detection networks for the realization of efficient search approaches and generalized object detection models in the future.

5.2 Generative adversarial networks

Generative adversarial networks (GAN) [203] is among the most promising types of unsupervised machine learning models, and foundational method in deep learning, particularly for synthetic data generation such as texts, images and music. GAN comprises two neural networks, generator and discriminator, competing for realistic data generation. The generator network creates unreal data such as audio, images, aiming to generate data that is identical to the real data. The duty of discriminator network is to differentiate the real data from the generator's unreal data. A probability score that indicates whether the data is real or unreal, is generated. The generator and discriminator networks are competitively and concurrently trained. While the discriminator's intelligence is being played upon by the generator, the discriminator attempts to accurately identify whether the input data is real or unreal. This process is referred to as a min-max game.

Specific loss functions are employed by GAN such as generator loss, which is penalized for any unreal generated data detected by the discriminator, and the discriminator loss, which is penalized for any unreal generated data undetected by the discriminator or vice-versa. There are a wide range of GAN applications such as image generation, image-to-image translation, super-resolution, data augmentation and style transfer. GAN's design exists in different variants, aiming to solve specific problems or improve performance. Conditional GAN (CGAN) [204] is a variant of GAN that allows data generation based on specific conditions. Wasserstein GANs (WGANs) [205] are a variant of GAN that improves the stability and training process by introducing a different loss function. Cycle-consistent GAN (CycleGAN) [206] is a variant of GAN employed for unpaired image-to-image translation tasks. GAN is a powerful means of producing excellent synthetic data; however, they have many challenges, particularly during training such as mode collapse, training instability and evaluation.

5.3 Vision-language models

Vision-language models (VLMs) are designed, as a class of machine learning models, for visual and textual information understanding and generation. Their design enables connection between CV and NLP, allowing processing and interpretation of images, videos and texts by the machines. Notable VLMs are Contrastive Language-Image Pretraining (CLIP) [207], ALIGN [208], VisualBERT [209], DALL·E 2 [210] and Foundational Language And Vision Alignment (FLAVA) [211]. VLMs are applied in areas such as autonomous vehicles, healthcare, assistive technologies, social media and e-commerce. However, VLMs have many challenges such as multimodal understanding, bias and fairness, scalability, and generalization. VLMs' capability for visual and textual information processing has significantly advanced AI. By these models, content (that involves both vision and language) can be understood and generated by machines. VLMs are leading in multimodal AI research, with applications across many industries.

5.4 Lightweight models

Lightweight object detection models are commonly used for object detection in images or videos. The models are optimized for computation, memory and speed efficiencies, without trading too much accuracy for the efficiencies. The main goal of lightweight object detection is to develop small-size, high speed, and real time-suitable object detection models, such as mobile devices, IoT devices embedded systems, edge devices. Often, lightweight object detection models address the limitations of traditional object detection models such as Faster R-CNN, YOLOv3, by abridging their architecture using techniques such as smaller backbone networks, quantization, pruning, knowledge distillation, single-shot detection models, Efficient Convolutional Operations and lightweight detection architectures.

Popular lightweight object detection models are YOLO, Tiny-DSOD [212], NanoDet [213], YOLObile [214], EfficientDet [142], EfficientNet [215], YOLOv12 [181], MobileNet-SSD and Tiny-YOLOv4. Lightweight object detection models are applied in different areas such as mobile devices, autonomous vehicles, drones, robotics, surveillance and security. However, lightweight object detection models have many challenges such as the tradeoff between accuracy and size, real-time and hardware constraints, and generalization. Lightweight object detection models provide a solution platform for performing real-time detection tasks in environments with resource constraints. Through the abovementioned techniques, computational efficiency and accuracy are balanced by the models, which are key to enabling a wide range of applications.

6. Conclusions

This chapter has provided a comprehensive review on object detection algorithms for digital imaging applications. The chapter reviews algorithms of deep learning models for object detection in 2D images in the recent years, focusing on their contributions and developmental trends from hand-crafted-based traditional methods to deep learning-based methods. This chapter divides the algorithms of the 2D object detection into anchor-based detection models, anchor-free detection models and Transformer-based detection models, covering detailed analysis of their contributions. Moreover, this chapter also presents the commonly used datasets and evaluation metrics for computer vision and object detection tasks. Also presented in this chapter are the research directions expected to address the challenges confronting the implementation and deployment of object detection models and smoothen the future of object detection. Currently, object detection research based on 2D images looks promising, and with the introduction of Transformer-based algorithms, research on algorithms for detecting both 2D and 3D objects in images will continue to soar.

Acknowledgements

The authors received funding from the Tshwane University of Technology, South Africa.

Conflict of interest

The authors declare no conflict of interest.

Author details

Rotimi-Williams Bello*, Pius A. Owolawi, Etienne A. van Wyk and Chunling Tu
Department of Computer Systems Engineering, Faculty of Information and Communication Technology, Tshwane University of Technology, South Africa

*Address all correspondence to: bellorw@tut.ac.za

References

[1] Bello RW, Mohamed ASA, Talib AZ. Contour extraction of individual cattle from an image using enhanced mask R-CNN instance segmentation method. IEEE Access. 2021;**9**:56984-57000

[2] Pietikäinen M. Local binary patterns. Scholarpedia. 2010;**5**(3):9775

[3] Lowe DG. Object recognition from local scale-invariant features. In: Proceedings of the Seventh IEEE International Conference on Computer Vision; Kerkyra (Corfu), Greece. Vol. 2. New York City: IEEE; 1999. pp. 1150-1157

[4] Lowe DG. Distinctive image features from scale-invariant keypoints. International Journal of Computer Vision. 2004;**60**:91-110

[5] Dalal N, Triggs B. Histograms of oriented gradients for human detection. In: 2005 IEEE Computer Society Conference on Computer Vision and Pattern Recognition(CVPR'05); San Diego, CA, USA. Vol. 1. New York City: IEEE; 2005. pp. 886-893

[6] Viola P, Jones M. Rapid object detection using a boosted cascade of simple features. In: Proceedings of the 2001 IEEE Computer Society Conference on Computer Vision and Pattern Recognition. CVPR 2001; Kauai, HI, USA. Vol. 1. New York City: IEEE; 2001. pp. 1-9

[7] Viola P, Jones MJ. Robust real-time face detection. International Journal of Computer Vision. 2004;**57**:137-154

[8] Krizhevsky A, Sutskever I, Hinton GE. Imagenet classification with deep convolutional neural networks. Advances in Neural Information Processing Systems. 2012;**25**:1-9

[9] Girshick R, Donahue J, Darrell T, Malik J. Rich feature hierarchies for accurate object detection and semantic segmentation. In: Proceedings of the IEEE Conference on Computer Vision and Pattern Recognition. New York City: IEEE; 2014. pp. 580-587

[10] Ren S, He K, Girshick R, Sun J. Faster R-CNN: Towards real-time object detection with region proposal networks. IEEE Transactions on Pattern Analysis and Machine Intelligence. 2016;**39**(6):1137-1149

[11] He K, Gkioxari G, Dollár P, Girshick R. Mask R-CNN. In: Proceedings of the IEEE International Conference on Computer Vision. New York City: IEEE; 2017. pp. 2961-2969

[12] Redmon J, Divvala S, Girshick R, Farhadi A. You only look once: Unified, real-time object detection. In: Proceedings of the IEEE Conference on Computer Vision and Pattern Recognition. New York City: IEEE; 2016. pp. 779-788

[13] Liu W, Anguelov D, Erhan D, Szegedy C, Reed S, Fu CY, et al. SSD: Single shot multibox detector. In: Computer Vision–ECCV 2016: 14th European Conference, Amsterdam, The Netherlands, October 11-14, 2016, Proceedings, Part I 14; Amsterdam, The Netherlands. Cham, Switzerland: Springer International Publishing; 2016. pp. 21-37

[14] Guo J, He H, He T, Lausen L, Li M, Lin H, et al. GluonCV and GluonNLP: Deep learning in computer vision and natural language processing. Journal of Machine Learning Research. 2020;**21**(23):1-7

[15] Carion N, Massa F, Synnaeve G, Usunier N, Kirillov A, Zagoruyko S.

End-to-end object detection with transformers. In: European Conference on Computer Vision. Cham: Springer International Publishing; 2020. pp. 213-229

[16] Vaswani A, Shazeer N, Parmar N, Uszkoreit J, Jones L, Gomez AN, et al. Attention is all you need. Advances in Neural Information Processing Systems. 2017;**30**:1-11

[17] He K, Zhang X, Ren S, Sun J. Deep residual learning for image recognition. In: Proceedings of the IEEE Conference on Computer Vision and Pattern Recognition. New York City: IEEE; 2016. pp. 770-778

[18] Liu Z, Lin Y, Cao Y, Hu H, Wei Y, Zhang Z, et al. Swin transformer: Hierarchical vision transformer using shifted windows. In: Proceedings of the IEEE/CVF International Conference on Computer Vision. New York City: IEEE; 2021. pp. 10012-10022

[19] Lin TY, Dollár P, Girshick R, He K, Hariharan B, Belongie S. Feature pyramid networks for object detection. In: Proceedings of the IEEE Conference on Computer Vision and Pattern Recognition. New York City: IEEE; 2017. pp. 2117-2125

[20] Chen Q, Wang Y, Yang T, Zhang X, Cheng J, Sun J. You only look one-level feature. In: Proceedings of the IEEE/CVF Conference on Computer Vision and Pattern Recognition. New York City: IEEE; 2021. pp. 13039-13048

[21] Dai J, Li Y, He K, Sun J. R-FCN: Object detection via region-based fully convolutional networks. Advances in Neural Information Processing Systems. 2016;**29**:1-9

[22] Liu Y, Zhang Y, Wang Y, Hou F, Yuan J, Tian J, et al. A survey of visual transformers. IEEE Transactions on Neural Networks and Learning Systems. 2023;**35**(6):7478-7498

[23] Suthaharan S, Suthaharan S. Support vector machine. In: Machine Learning Models and Algorithms for Big Data Classification. Integrated Series in Information Systems. Vol. 36. Boston, MA: Springer; 2016. pp. 207-235

[24] Mangasarian OL, Musicant DR. Lagrangian support vector machines. Journal of Machine Learning Research. 2001;**1**(Mar):161-177

[25] Tsochantaridis I, Joachims T, Hofmann T, Altun Y. Large margin methods for structured and interdependent output variables. Journal of Machine Learning Research. 2005;**6**(Sep):1453-1484

[26] Blaschko MB, Lampert CH. Learning to localize objects with structured output regression. In: Computer Vision–ECCV 2008: 10th European Conference on Computer Vision, Marseille, France, October 12-18, 2008, Proceedings, Part I 10. Berlin Heidelberg: Springer; 2008. pp. 2-15

[27] Schapire RE. Explaining AdaBoost. In: Empirical Inference: Festschrift in Honor of Vladimir N. Vapnik. Berlin, Heidelberg: Springer Berlin Heidelberg; 2013. pp. 37-52

[28] Felzenszwalb P, McAllester D, Ramanan D. A discriminatively trained, multiscale, deformable part model. In: 2008 IEEE Conference on Computer Vision and Pattern Recognition; Anchorage, AK, USA. New York City: IEEE; 2008. pp. 1-8

[29] Van De Sande K, Gevers T, Snoek C. Evaluating color descriptors for object and scene recognition. IEEE Transactions on Pattern Analysis and Machine Intelligence. 2009;**32**(9):1582-1596

[30] Ke Y, Sukthankar R. PCA-SIFT: A more distinctive representation for local image descriptors. In: Proceedings of the 2004 IEEE Computer Society Conference on Computer Vision and Pattern Recognition, 2004. CVPR 2004; Washington, DC, USA. Vol. 2. New York City: IEEE; 2004. pp. 1-9

[31] Lienhart R, Maydt J. An extended set of HAAR-like features for rapid object detection. In: Proceedings International Conference on Image Processing; Rochester, NY, USA. Vol. 1. New York City: IEEE; 2002. pp. 1-4

[32] Barinova O, Lempitsky V, Kholi P. On detection of multiple object instances using hough transforms. IEEE Transactions on Pattern Analysis and Machine Intelligence. 2012;**34**(9):1773-1784

[33] Uijlings JR, Van De Sande KE, Gevers T, Smeulders AW. Selective search for object recognition. International Journal of Computer Vision. 2013;**104**:154-171

[34] He K, Zhang X, Ren S, Sun J. Spatial pyramid pooling in deep convolutional networks for visual recognition. IEEE Transactions on Pattern Analysis and Machine Intelligence. 2015;**37**(9):1904-1916

[35] Lazebnik S, Schmid C, Ponce J. Beyond bags of features: Spatial pyramid matching for recognizing natural scene categories. In: 2006 IEEE Computer Society Conference on Computer Vision and Pattern Recognition (CVPR'06); New York, NY, USA. Vol. 2. New York City: IEEE; 2006. pp. 2169-2178

[36] Girshick R. Fast R-CNN. In: Proceedings of the IEEE International Conference on Computer Vision. New York City: IEEE; 2015. pp. 1440-1448

[37] Long J, Shelhamer E, Darrell T. Fully convolutional networks for semantic segmentation. In: Proceedings of the IEEE Conference on Computer Vision and Pattern Recognition. New York City: IEEE; 2015. pp. 3431-3440

[38] Li Z, Chen Y, Yu G, Deng Y. R-FCN++: Towards accurate region-based fully convolutional networks for object detection. In: Proceedings of the AAAI Conference on Artificial Intelligence. Vol. 32, No. 1. Palo Alto, California, USA: Association for the Advancement of Artificial Intelligence (AAAI) Press; 2018

[39] Dai J, Qi H, Xiong Y, Li Y, Zhang G, Hu H, et al. Deformable convolutional networks. In: Proceedings of the IEEE International Conference on Computer Vision. New York City: IEEE; 2017. pp. 764-773

[40] Li Z, Peng C, Yu G, Zhang X, Deng Y, Sun J. Light-head R-CNN: In defense of two-stage object detector. arXiv preprint. 2017;**2**:1-9. arXiv preprint arXiv:1711.07264

[41] Bello RW, Mohamed ASA, Talib AZ. Enhanced mask R-CNN for herd segmentation. International Journal of Agricultural and Biological Engineering. 2021;**14**(4):238-244

[42] Bello RW, Owolawi PA, van Wyk EA, Tu C. Transfer learning-driven cattle instance segmentation using deep learning models. Agriculture. 2024;**14**(12):2282

[43] Bello R-W, Owolawi PA, van Wyk AE, Tu C. Cattle Instance Segmentation by Transfer Learning Approach Using Deep Learning Models for Sustainable Livestock Farming. London, UK: IntechOpen; 2025. pp. 1-17. DOI: 10.5772/intechopen.1009155

[44] Kim SW, Kook HK, Sun JY, Kang MC, Ko SJ. Parallel feature pyramid network for object detection. In: Proceedings of the European Conference on Computer Vision (ECCV). Cham, Switzerland: Springer International Publishing; 2018. pp. 234-250

[45] Hu M, Li Y, Fang L, Wang S. A2-FPN: Attention aggregation based feature pyramid network for instance segmentation. In: Proceedings of the IEEE/CVF Conference on Computer Vision and Pattern Recognition. New York City: IEEE; 2021. pp. 15343-15352

[46] Dai J, He K, Sun J. Instance-aware semantic segmentation via multi-task network cascades. In: Proceedings of the IEEE Conference on Computer Vision and Pattern Recognition. New York City: IEEE; 2016. pp. 3150-3158

[47] Cai Z, Vasconcelos N. Cascade R-CNN: Delving into high quality object detection. In: Proceedings of the IEEE Conference on Computer Vision and Pattern Recognition. New York City: IEEE; 2018. pp. 6154-6162

[48] Gidaris S, Komodakis N. Object detection via a multi-region and semantic segmentation-aware CNN model. In: Proceedings of the IEEE International Conference on Computer Vision. New York City: IEEE; 2015. pp. 1134-1142

[49] Cai Z, Fan Q, Feris RS, Vasconcelos N. A unified multi-scale deep convolutional neural network for fast object detection. In: Computer Vision–ECCV 2016: 14th European Conference, Amsterdam, The Netherlands, October 11-14, 2016, Proceedings, Part IV 14; Amsterdam, The Netherlands. Cham, Switzerland: Springer International Publishing; 2016. pp. 354-370

[50] Wang X, Shrivastava A, Gupta A. A-fast-RCNN: Hard positive generation via adversary for object detection. In: Proceedings of the IEEE Conference on Computer Vision and Pattern Recognition. New York City: IEEE; 2017. pp. 2606-2615

[51] Pang J, Chen K, Shi J, Feng H, Ouyang W, Lin D. Libra R-CNN: Towards balanced learning for object detection. In: Proceedings of the IEEE/CVF Conference on Computer Vision and Pattern Recognition. New York City: IEEE; 2019. pp. 821-830

[52] Sun P, Zhang R, Jiang Y, Kong T, Xu C, Zhan W, et al. Sparse R-CNN: End-to-end object detection with learnable proposals. In: Proceedings of the IEEE/CVF Conference on Computer Vision and Pattern Recognition. New York City: IEEE; 2021. pp. 14454-14463

[53] Lee H, Eum S, Kwon H. ME R-CNN: Multi-expert R-CNN for object detection. IEEE Transactions on Image Processing. 2019;**29**:1030-1044

[54] Redmon J, Farhadi A. YOLO9000: better, faster, stronger. In: Proceedings of the IEEE Conference on Computer Vision and Pattern Recognition. New York City: IEEE; 2017. pp. 7263-7271

[55] Redmon J, Farhadi A. YOLOv3: An incremental improvement. arXiv preprint. 2018;**1**:1-6. arXiv preprint arXiv:1804.02767

[56] Bochkovskiy A, Wang CY, Liao HYM. YOLOv4: Optimal speed and accuracy of object detection. arXiv preprint. 2020;**1**:1-17. arXiv preprint arXiv:2004.10934

[57] Mahasin M, Dewi IA. Comparison of CSPDarkNet53, CSPResNeXt-50, and EfficientNet-B0 backbones on YOLOv4 as object detector. International Journal of Engineering, Science and Information Technology. 2022;**2**(3):64-72

[58] Yun S, Han D, Oh SJ, Chun S, Choe J, Yoo Y. CutMix: Regularization strategy to train strong classifiers with localizable features. In: Proceedings of the IEEE/CVF International Conference on Computer Vision, 6032. New York City: IEEE; 2019. p. 6023

[59] Liu S, Qi L, Qin H, Shi J, Jia J. Path aggregation network for instance segmentation. In: Proceedings of the IEEE Conference on Computer Vision and Pattern Recognition. New York City: IEEE; 2018. pp. 8759-8768

[60] Jocher G, Stoken A, Borovec J, Changyu L, Hogan A, Diaconu L, et al. Ultralytics/YOLOv5: v3. 1-bug fixes and performance improvements. 2020. DOI: 10.5281/zenodo.4154370

[61] Zhang H, Cisse M, Dauphin YN, Lopez-Paz D. Mixup: Beyond empirical risk minimization. arXiv preprint. 2017;**2**:1-13. arXiv preprint arXiv:1710.09412

[62] Ghiasi G, Cui Y, Srinivas A, Qian R, Lin TY, Cubuk ED, et al. Simple copy-paste is a strong data augmentation method for instance segmentation. In: Proceedings of the IEEE/CVF Conference on Computer Vision and Pattern Recognition. New York City: IEEE; 2021. pp. 2918-2928

[63] Buslaev A, Iglovikov VI, Khvedchenya E, Parinov A, Druzhinin M, Kalinin AA. Albumentations: Fast and flexible image augmentations. Information. 2020;**11**(2):1-20

[64] Li C, Li L, Jiang H, Weng K, Geng Y, Li L, et al. YOLOv6: A single-stage object detection framework for industrial applications. arXiv preprint. 2022;**2**:1-17. arXiv preprint arXiv:2209.02976

[65] Ding X, Zhang X, Ma N, Han J, Ding G, Sun J. REPVGG: Making VGG-style convnets great again. In: Proceedings of the IEEE/CVF Conference on Computer Vision and Pattern Recognition. New York City: IEEE; 2021. pp. 13733-13742

[66] Wang CY, Liao HYM, Wu YH, Chen PY, Hsieh JW, Yeh IH. CSPNet: A new backbone that can enhance learning capability of CNN. In: Proceedings of the IEEE/CVF Conference on Computer Vision and Pattern Recognition Workshops. New York City: IEEE; 2020. pp. 390-391

[67] Rezatofighi H, Tsoi N, Gwak J, Sadeghian A, Reid I, Savarese S. Generalized intersection over union: A metric and a loss for bounding box regression. In: Proceedings of the IEEE/CVF Conference on Computer Vision and Pattern Recognition. New York City: IEEE; 2019. pp. 658-666

[68] Gevorgyan Z. SIoU loss: More powerful learning for bounding box regression. arXiv preprint. 2022;**1**:1-12. arXiv preprint arXiv:2205.12740

[69] Wang CY, Bochkovskiy A, Liao HYM. YOLOv7: Trainable bag-of-freebies sets new state-of-the-art for real-time object detectors. In: Proceedings of the IEEE/CVF Conference on Computer Vision and Pattern Recognition. New York City: IEEE; 2023. pp. 7464-7475

[70] Hussain M. YOLOv1 to v8: Unveiling each variant–A comprehensive review of yolo. IEEE Access. 2024;**12**:42816-42833

[71] Lee CY, Xie S, Gallagher P, Zhang Z, Tu Z. Deeply-supervised nets. In: Artificial Intelligence and Statistics; San Diego, CA, USA. Cambridge, MA, USA: PMLR; 2015. pp. 562-570

[72] Jocher G, Chaurasia A, Qiu J. Ultralytics YOLOv8. 2023. Available from: https://docs.ultralytics.com/

[73] Li H, Wu A, Jiang Z, Liu F, Luo M. Improving object detection in YOLOv8n with the C2f-f module and multi-scale fusion reconstruction. In: 2024 IEEE 6th Advanced Information Management, Communicates, Electronic and Automation Control Conference (IMCEC); Chongqing, China. Vol. 6. New York City: IEEE; 2024. pp. 374-379

[74] Zheng Z, Wang P, Liu W, Li J, Ye R, Ren D. Distance-IoU loss: Faster and better learning for bounding box regression. In: Proceedings of the AAAI Conference on Artificial Intelligence. Vol. 34, No. 07. Palo Alto, California USA: Association for the Advancement of Artificial Intelligence (AAAI) Press; 2020. pp. 12993-13000

[75] Li X, Wang W, Wu L, Chen S, Hu X, Li J, et al. Generalized focal loss: Learning qualified and distributed bounding boxes for dense object detection. Advances in Neural Information Processing Systems. 2020;**33**:21002-21012

[76] Wang CY, Yeh IH, Mark Liao HY. YOLOv9: Learning what you want to learn using programmable gradient information. In: European Conference on Computer Vision. Cham, Switzerland: Springer Nature; 2024. pp. 1-21

[77] Lin Z, Wang Y, Zhang J, Chu X. DynamicDet: A unified dynamic architecture for object detection. In: Proceedings of the IEEE/CVF Conference on Computer Vision and Pattern Recognition. New York City: IEEE; 2023. pp. 6282-6291

[78] Wang CY, Liao HYM, Yeh IH. Designing network design strategies through gradient path analysis. arXiv preprint. 2022;**1**:1-12. arXiv preprint arXiv:2211.04800

[79] Wang A, Chen H, Liu L, Chen K, Lin Z, Han J. YOLOv10: Real-time end-to-end object detection. Advances in Neural Information Processing Systems. 2024;**37**:107984-108011

[80] Nguyen DMT, Huynh-The T. RS-YOLOv10: Enhancing YOLOv10 for accurate small-object detection. In: 2025 19th International Conference on Ubiquitous Information Management and Communication (IMCOM); Bangkok, Thailand. New York City: IEEE; 2025. pp. 1-7

[81] Chollet F. Xception: Deep learning with depthwise separable convolutions. In: Proceedings of the IEEE Conference on Computer Vision and Pattern Recognition. New York City: IEEE; 2017. pp. 1251-1258

[82] Jocher G, Qiu J. Ultralytics YOLOv11. 2024. Available from: https://docs.ultralytics.com/models/yolo11/

[83] Long X, Deng K, Wang G, Zhang Y, Dang Q, Gao Y, et al. PP-YOLO: An effective and efficient implementation of object detector. arXiv preprint. 2020;**3**:1-8. arXiv preprint arXiv:2007.12099

[84] Ma Y, Yu D, Wu T, Wang H. PaddlePaddle: An open-source deep learning platform from industrial practice. Frontiers of Data and Domputing. 2019;**1**(1):105-115

[85] Huang X, Wang X, Lv W, Bai X, Long X, Deng K, et al. PP-YOLOv2: A practical object detector. arXiv preprint. 2021;**1**:1-7. arXiv preprint arXiv:2104.10419

[86] Xu S, Wang X, Lv W, Chang Q, Cui C, Deng K, et al. PP-YOLOE: An evolved version of YOLO. 2022: 1-7. arXiv preprint arXiv:2203.16250

[87] Wang X, Wang G, Dang Q, Liu Y, Hu X, Yu D. PP-YOLOE-R: An efficient anchor-free rotated object detector.

arXiv preprint. 2022;**1**:1-6. arXiv preprint arXiv:2211.02386

[88] Wang CY, Yeh IH, Liao HYM. You only learn one representation: Unified network for multiple tasks. arXiv preprint. 2021;**1**:1-11. arXiv preprint arXiv:2105.04206

[89] Ge Z, Liu S, Wang F, Li Z, Sun J. YOLOX: Exceeding YOLO series in 2021. arXiv preprint. 2021;**2**:1-7. arXiv preprint arXiv:2107.08430

[90] Xu X, Jiang Y, Chen W, Huang Y, Zhang Y, Sun X. DAMO-YOLO: A report on real-time object detection design. arXiv preprint. 2022;**4**:1-10. arXiv preprint arXiv:2211.15444

[91] Wu Z, Zou X, Zhou W, Huang J. YOLOX-PAI: An improved YOLOX, stronger and faster than YOLOv6. arXiv preprint. 2022;**3**:1-5. arXiv preprint arXiv:2208.13040

[92] Wang C, He W, Nie Y, Guo J, Liu C, Wang Y, et al. Gold-YOLO: Efficient object detector via gather-and-distribute mechanism. Advances in Neural Information Processing Systems. 2023;**36**:51094-51112

[93] Chen Y, Yuan X, Wang J, Wu R, Li X, Hou Q, et al. YOLO-MS: Rethinking multi-scale representation learning for real-time object detection. IEEE Transactions on Pattern Analysis and Machine Intelligence. 2025:1-13

[94] Huang L, Li W, Tan Y, Shen L, Yu J, Fu H. YOLOCS: Object detection based on dense channel compression for feature spatial solidification. Knowledge-Based Systems. 2025;**113024**:1-25

[95] Zhang Z, Lu X, Cao G, Yang Y, Jiao L, Liu F. ViT-YOLO: Transformer-based YOLO for object detection. In: Proceedings of the IEEE/CVF International Conference on Computer Vision. New York City: IEEE; 2021. pp. 2799-2808

[96] Guo Z, Wang C, Yang G, Huang Z, Li G. MSFT-YOLO: Improved YOLOv5 based on transformer for detecting defects of steel surface. Sensors. 2022;**22**(9):1-15

[97] Liu Y, He G, Wang Z, Li W, Huang H. NRT-YOLO: Improved YOLOv5 based on nested residual transformer for tiny remote sensing object detection. Sensors. 2022;**22**(13):1-16

[98] Wang S, Gao S, Zhou L, Liu R, Zhang H, Liu J, et al. YOLO-SD: Small ship detection in SAR images by multi-scale convolution and feature transformer module. Remote Sensing. 2022;**14**(20):1-21

[99] Ouyang H. DEYO: DETR with YOLO for step-by-step object detection. arXiv preprint. 2022;**3**:1-10. arXiv preprint arXiv:2211.06588

[100] Yang F, Huang L, Tan X, Yuan Y. FasterNet-SSD: A small object detection method based on SSD model. Signal, Image and Video Processing. 2024;**18**(1):173-180

[101] Fu CY, Liu W, Ranga A, Tyagi A, Berg AC. DSSD: Deconvolutional single shot detector. arXiv preprint. 2017;**1**:1-11. arXiv preprint arXiv:1701.06659

[102] Jeong J, Park H, Kwak N. Enhancement of SSD by concatenating feature maps for object detection. arXiv preprint. 2017;**1**:1-12. arXiv preprint arXiv:1705.09587

[103] Zheng L, Fu C, Zhao Y. Extend the shallow part of single shot multibox detector via convolutional neural network. In: Tenth International Conference on Digital Image Processing

(ICDIP 2018); Shanghai, China. Vol. 10806. Bellingham, Washington, USA: SPIE; 2018. pp. 287-293

[104] Cao G, Xie X, Yang W, Liao Q, Shi G, Wu J. Feature-fused SSD: Fast detection for small objects. In: Ninth International Conference on Graphic and Image Processing (ICGIP 2017); Qingdao, China. Vol. 10615. Bellingham, Washington, USA: SPIE; 2018. pp. 381-388

[105] Yang J, Wang L. Feature fusion and enhancement for single shot multibox detector. In: 2019 Chinese Automation Congress (CAC); Hangzhou, China. New York City: IEEE; 2019. pp. 2766-2770

[106] Yi J, Wu P, Metaxas DN. ASSD: Attentive single shot multibox detector. Computer Vision and Image Understanding. 2019;**189**(102827):1-9

[107] Chandio A, Gui G, Kumar T, Ullah I, Ranjbarzadeh R, Roy AM, et al. Precise single-stage detector. arXiv preprint. 2022;**1**:1-33. arXiv preprint arXiv:2210.04252

[108] Lin TY, Goyal P, Girshick R, He K, Dollár P. Focal loss for dense object detection. In: Proceedings of the IEEE International Conference on Computer Vision. New York City: IEEE; 2017. pp. 2980-2988

[109] Fu CY, Shvets M, Berg AC. RetinaMask: Learning to predict masks improves state-of-the-art single-shot detection for free. arXiv preprint. 2019;**1**:1-11. arXiv preprint arXiv:1901.03353

[110] Jaeger PF, Kohl SA, Bickelhaupt S, Isensee F, Kuder TA, Schlemmer HP, et al. Retina U-net: Embarrassingly simple exploitation of segmentation supervision for medical object detection. In: Machine Learning for Health Workshop; Virtual Conference, at NeurIPS. Cambridge, MA, USA: PMLR; 2020. pp. 171-183

[111] Ronneberger O, Fischer P, Brox T. U-net: Convolutional networks for biomedical image segmentation. In: Medical Image Computing and Computer-Assisted Intervention–MICCAI 2015: 18th International Conference, Munich, Germany, October 5-9, 2015, Proceedings, Part III 18; Munich, Germany. Cham, Switzerland: Springer International Publishing; 2015. pp. 234-241

[112] Du X, Zoph B, Hung WC, Lin TY. Simple training strategies and model scaling for object detection. arXiv preprint. 2021;**1**:1-9. arXiv preprint arXiv:2107.00057

[113] Zhang S, Wen L, Bian X, Lei Z, Li SZ. Single-shot refinement neural network for object detection. In: Proceedings of the IEEE Conference on Computer Vision and Pattern Recognition. New York City: IEEE; 2018. pp. 4203-4212

[114] Nie J, Anwer RM, Cholakkal H, Khan FS, Pang Y, Shao L. Enriched feature guided refinement network for object detection. In: Proceedings of the IEEE/CVF International Conference on Computer Vision. New York City: IEEE; 2019. pp. 9537-9546

[115] Chen W, Li Y, Tian Z, Zhang F. 2D and 3D object detection algorithms from images: A survey. Array. 2023;**19**:1-23

[116] Law H, Deng J. CornerNet: Detecting objects as paired keypoints. In: Proceedings of the European Conference on Computer Vision (ECCV). Cham, Switzerland: Springer International Publishing; 2018. pp. 734-750

[117] Law H, Teng Y, Russakovsky O, Deng J. CornerNet-lite: Efficient

keypoint based object detection. arXiv preprint. 2019;**2**:1-15. arXiv preprint arXiv:1904.08900

[118] Rashwan A, Kalra A, Poupart P. Matrix nets: A new deep architecture for object detection. In: Proceedings of the IEEE/CVF International Conference on Computer Vision Workshops. New York City: IEEE; 2019. pp. 1-4

[119] Duan K, Bai S, Xie L, Qi H, Huang Q, Tian Q. CenterNet: Keypoint triplets for object detection. In: Proceedings of the IEEE/CVF International Conference on Computer Vision. New York City: IEEE; 2019. pp. 6569-6578

[120] Zhou X, Zhuo J, Krahenbuhl P. Bottom-up object detection by grouping extreme and center points. In: Proceedings of the IEEE/CVF Conference on Computer Vision and Pattern Recognition. New York City: IEEE; 2019. pp. 850-859

[121] Yoo D, Park S, Lee JY, Paek AS, So Kweon I. AttentionNet: Aggregating weak directions for accurate object detection. In: Proceedings of the IEEE International Conference on Computer Vision. New York City: IEEE; 2015. pp. 2659-2667

[122] Wang X, Chen K, Huang Z, Yao C, Liu W. Point linking network for object detection. arXiv preprint. 2017;**2**:1-10. arXiv preprint arXiv:1706.03646

[123] Dong Z, Li G, Liao Y, Wang F, Ren P, Qian C. CentripetalNet: Pursuing high-quality keypoint pairs for object detection. In: Proceedings of the IEEE/CVF Conference on Computer Vision and Pattern Recognition. New York City: IEEE; 2020. pp. 10519-10528

[124] Tychsen-Smith L, Petersson L. DeNet: Scalable real-time object detection with directed sparse sampling. In: Proceedings of the IEEE International Conference on Computer Vision. New York City: IEEE; 2017. pp. 428-436

[125] Lu X, Li B, Yue Y, Li Q, Yan J. Grid R-CNN. In: Proceedings of the IEEE/CVF Conference on Computer Vision and Pattern Recognition. New York City: IEEE; 2019. pp. 7363-7372

[126] Yang Z, Liu S, Hu H, Wang L, Lin S. RepPoints: Point set representation for object detection. In: Proceedings of the IEEE/CVF International Conference on Computer Vision. New York City: IEEE; 2019. pp. 9657-9666

[127] Huang L, Yang Y, Deng Y, Yu Y. Densebox: Unifying landmark localization with end to end object detection. arXiv preprint. 2015;**3**:1-13. arXiv preprint arXiv:1509.04874

[128] Tian Z, Shen C, Chen H, He T. FCOS: Fully convolutional one-stage object detection. In: Proceedings of the IEEE/CVF International Conference on Computer Vision. New York City: IEEE; 2019. pp. 9627-9636

[129] Zhu C, He Y, Savvides M. Feature selective anchor-free module for single-shot object detection. In: Proceedings of the IEEE/CVF Conference on Computer Vision and Pattern Recognition. New York City: IEEE; 2019. pp. 840-849

[130] Tian Z, Shen C, Chen H, He T. FCOS: A simple and strong anchor-free object detector. IEEE Transactions on Pattern Analysis and Machine Intelligence. 2020;**44**(4):1922-1933

[131] Qiu H, Ma Y, Li Z, Liu S, Sun J. Borderdet: Border feature for dense object detection. In: European Conference on Computer Vision. Cham: Springer International Publishing; 2020. pp. 549-564

[132] Lee Y, Park J. CenterMask: Real-time anchor-free instance segmentation. In: Proceedings of the IEEE/CVF Conference on Computer Vision and Pattern Recognition. New York City: IEEE; 2020. pp. 13906-13915

[133] Kong T, Sun F, Liu H, Jiang Y, Li L, Shi J. FoveaBox: Beyound anchor-based object detection. IEEE Transactions on Image Processing. 2020;**29**:7389-7398

[134] Zhu C, Chen F, Shen Z, Savvides M. Soft anchor-point object detection. In: Computer Vision–ECCV 2020: 16th European Conference, Glasgow, UK, August 23-28, 2020, Proceedings, Part IX 16; Virtual Conference. Cham, Switzerland: Springer International Publishing; 2020. pp. 91-107

[135] Qiu H, Li H, Wu Q, Cui J, Song Z, Wang L, et al. CrossDet: Crossline representation for object detection. In: Proceedings of the IEEE/CVF International Conference on Computer Vision. New York City: IEEE; 2021. pp. 3195-3204

[136] Yu G, Chang Q, Lv W, Xu C, Cui C, Ji W, et al. PP-PicoDet: A better real-time object detector on mobile devices. arXiv preprint. 2021;**1**:1-9. arXiv preprint arXiv:2111.00902

[137] Najibi M, Rastegari M, Davis LS. G-CNN: An iterative grid based object detector. In: Proceedings of the IEEE Conference on Computer Vision and Pattern Recognition. New York City: IEEE; 2016. pp. 2369-2377

[138] Kong T, Sun F, Yao A, Liu H, Lu M, Chen Y. RON: Reverse connection with objectness prior networks for object detection. In: Proceedings of the IEEE Conference on Computer Vision and Pattern Recognition. New York City: IEEE; 2017. pp. 5936-5944

[139] Zhou P, Ni B, Geng C, Hu J, Xu Y. Scale-transferable object detection. In: Proceedings of the IEEE Conference on Computer Vision and Pattern Recognition. New York City: IEEE; 2018. pp. 528-537

[140] Cao J, Pang Y, Li X. Triply supervised decoder networks for joint detection and segmentation. In: Proceedings of the IEEE/CVF Conference on Computer Vision and Pattern Recognition. New York City: IEEE; 2019. pp. 7392-7401

[141] Chen X, Yu J, Kong S, Wu Z, Wen L. Joint anchor-feature refinement for real-time accurate object detection in images and videos. IEEE Transactions on Circuits and Systems for Video Technology. 2020;**31**(2):594-607

[142] Tan M, Pang R, Le QV. EfficientDet: Scalable and efficient object detection. In: Proceedings of the IEEE/CVF Conference on Computer Vision and Pattern Recognition. New York City: IEEE; 2020. pp. 10781-10790

[143] Wang S, Gong Y, Xing J, Huang L, Huang C, Hu W. RDSNET: A new deep architecture forreciprocal object detection and instance segmentation. In: Proceedings of the AAAI Conference on Artificial Intelligence. Vol. 34, No. 07. Palo Alto, California USA: Association for the Advancement of Artificial Intelligence (AAAI) Press; 2020. pp. 12208-12215

[144] Feng C, Zhong Y, Gao Y, Scott MR, Huang W. TOOD: Task-aligned one-stage object detection. In: 2021 IEEE/CVF International Conference on Computer Vision (ICCV); Virtual Conference. New York City: IEEE Computer Society; 2021. pp. 3490-3499

[145] Szegedy C, Toshev A, Erhan D. Deep neural networks for object

detection. Advances in Neural Information Processing Systems. 2013;**26**:1-9

[146] Sermanet P, Eigen D, Zhang X, Mathieu M, Fergus R, LeCun Y. OverFeat: Integrated recognition, localization and detection using convolutional networks. arXiv preprint. 2013;**4**:1-16. arXiv preprint arXiv:1312.6229

[147] Kuhn HW. The Hungarian method for the assignment problem. Naval Research Logistics Quarterly. 1955;**2**(1-2):83-97

[148] Zheng M, Gao P, Zhang R, Li K, Wang X, Li H, et al. End-to-end object detection with adaptive clustering transformer. arXiv preprint. 2020;**2**:1-14. arXiv preprint arXiv:2011.09315

[149] Gao P, Zheng M, Wang X, Dai J, Li H. Fast convergence of DETR with spatially modulated co-attention. In: Proceedings of the IEEE/CVF International Conference on Computer Vision. New York City: IEEE; 2021. pp. 3621-3630

[150] Dai X, Chen Y, Yang J, Zhang P, Yuan L, Zhang L. Dynamic DETR: End-to-end object detection with dynamic attention. In: Proceedings of the IEEE/CVF International Conference on Computer Vision. New York City: IEEE; 2021. pp. 2988-2997

[151] Zhu X, Su W, Lu L, Li B, Wang X, Dai J. Deformable DETR: Deformable transformers for end-to-end object detection. arXiv preprint. 2020;**4**:1-16. arXiv preprint arXiv:2010.04159

[152] Roh B, Shin J, Shin W, Kim S. Sparse DETR: Efficient end-to-end object detection with learnable sparsity. arXiv preprint. 2021;**2**:1-23. arXiv preprint arXiv:2111.14330

[153] Liu S, Li F, Zhang H, Yang X, Qi X, Su H, et al. DAB-DETR: Dynamic anchor boxes are better queries for DETR. arXiv preprint. 2022;**4**:1-19. arXiv preprint arXiv:2201.12329

[154] Meng D, Chen X, Fan Z, Zeng G, Li H, Yuan Y, et al. Conditional DETR for fast training convergence. In: Proceedings of the IEEE/CVF International Conference on Computer Vision. New York City: IEEE; 2021. pp. 3651-3660

[155] Wang Y, Zhang X, Yang T, Sun J. Anchor DETR: Query design for transformer-based detector. In: Proceedings of the AAAI Conference on Artificial Intelligence. Vol. 36, No. 3. Palo Alto, California USA: Association for the Advancement of Artificial Intelligence (AAAI) Press; 2022. pp. 2567-2575

[156] Li F, Zhang H, Liu S, Guo J, Ni LM, Zhang L. DN-DETR: Accelerate DETR training by introducing query denoising. In: Proceedings of the IEEE/CVF Conference on Computer Vision and Pattern Recognition. New York City: IEEE; 2022. pp. 13619-13627

[157] Zhang H, Li F, Liu S, Zhang L, Su H, Zhu J, et al. DINO: DETR with improved denoising anchor boxes for end-to-end object detection. arXiv preprint. 2022;**4**:1-23. arXiv preprint arXiv:2203.03605

[158] Dai Z, Cai B, Lin Y, Chen J. UP-DETR: Unsupervised pre-training for object detection with transformers. In: Proceedings of the IEEE/CVF Conference on Computer Vision and Pattern Recognition. New York City: IEEE; 2021. pp. 1601-1610

[159] Yao Z, Ai J, Li B, Zhang C. Efficient DETR: Improving end-to-end object detector with dense prior. arXiv

preprint. 2021;**1**:1-10. arXiv preprint arXiv:2104.01318

[160] Sun Z, Cao S, Yang Y, Kitani KM. Rethinking transformer-based set prediction for object detection. In: Proceedings of the IEEE/CVF International Conference on Computer Vision. New York City: IEEE; 2021. pp. 3611-3620

[161] Liu F, Wei H, Zhao W, Li G, Peng J, Li Z. WB-DETR: Transformer-based detector without backbone. In: Proceedings of the IEEE/CVF International Conference on Computer Vision. New York City: IEEE; 2021. pp. 2979-2987

[162] Wang T, Yuan L, Chen Y, Feng J, Yan S. PNP-DETR: Towards efficient visual analysis with transformers. In: Proceedings of the IEEE/CVF International Conference on Computer Vision. New York City: IEEE; 2021. pp. 4661-4670

[163] Lin J, Mao X, Chen Y, Xu L, He Y, Xue H. D2ETR: Decoder-only DETR with computationally efficient cross-scale attention. arXiv preprint. 2022;**1**:1-18. arXiv preprint arXiv:2203.00860

[164] Wang W, Cao Y, Zhang J, Tao D. FP-DETR: Detection transformer advanced by fully pre-training. In: International Conference on Learning Representations. NY, United States: Curran Associates, Inc.; 2021. pp. 1-14

[165] Cao X, Yuan P, Feng B, Niu K. CF-DETR: Coarse-to-fine transformers for end-to-end object detection. In: Proceedings of the AAAI Conference on Artificial Intelligence. Vol. 36, No. 1. Palo Alto, California, USA: Association for the Advancement of Artificial Intelligence (AAAI) Press; 2022. pp. 185-193

[166] Chen Z, Zhang J, Tao D. Recurrent glimpse-based decoder for detection with transformer. In: Proceedings of the IEEE/CVF Conference on Computer Vision and Pattern Recognition. New York City: IEEE; 2022. pp. 5260-5269

[167] Zong Z, Song G, Liu Y. DETRS with collaborative hybrid assignments training. In: Proceedings of the IEEE/CVF International Conference on Computer Vision. New York City: IEEE; 2023. pp. 6748-6758

[168] Kamath A, Singh M, LeCun Y, Synnaeve G, Misra I, Carion N. MDETR-modulated detection for end-to-end multi-modal understanding. In: Proceedings of the IEEE/CVF International Conference on Computer Vision. New York City: IEEE; 2021. pp. 1780-1790

[169] Chen Y, Chen Q, Sun P, Chen S, Wang J, Cheng J. Enhancing your trained DETRS with box refinement. arXiv preprint. 2023;**1**:1-13. arXiv preprint arXiv:2307.11828

[170] Dosovitskiy A, Beyer L, Kolesnikov A, Weissenborn D, Zhai X, Unterthiner T, et al. An image is worth 16x16 words: Transformers for image recognition at scale. arXiv preprint. 2020;**2**:1-22. arXiv preprint arXiv:2010.11929

[171] Beal J, Kim E, Tzeng E, Park DH, Zhai A, Kislyuk D. Toward transformer-based object detection. arXiv preprint. 2020;**1**:1-11. arXiv preprint arXiv:2012.09958

[172] Wang W, Xie E, Li X, Fan DP, Song K, Liang D, et al. Pyramid vision transformer: A versatile backbone for dense prediction without convolutions. In: Proceedings of the IEEE/CVF International Conference on Computer Vision. New York City: IEEE; 2021. pp. 568-578

[173] Wang W, Xie E, Li X, Fan DP, Song K, Liang D, et al. PVT V2: Improved baselines with pyramid vision transformer. Computational Visual Media. 2022;**8**(3):415-424

[174] Liu Z, Hu H, Lin Y, Yao Z, Xie Z, Wei Y, et al. SWIN transformer V2: Scaling up capacity and resolution. In: Proceedings of the IEEE/CVF Conference on Computer Vision and Pattern Recognition. New York City: IEEE; 2022. pp. 12009-12019

[175] Li Y, Mao H, Girshick R, He K. Exploring plain vision transformer backbones for object detection. In: European Conference on Computer Vision. Springer Nature Switzerland: Cham; 2022. pp. 280-296

[176] He K, Chen X, Xie S, Li Y, Dollár P, Girshick R. Masked autoencoders are scalable vision learners. In: Proceedings of the IEEE/CVF Conference on Computer Vision and Pattern Recognition. New York City: IEEE; 2022. pp. 16000-16009

[177] Fang Y, Liao B, Wang X, Fang J, Qi J, Wu R, et al. You only look at one sequence: Rethinking transformer in vision through object detection. Advances in Neural Information Processing Systems. 2021;**34**:26183-26197

[178] Fan H, Xiong B, Mangalam K, Li Y, Yan Z, Malik J, et al. Multiscale vision transformers. In: Proceedings of the IEEE/CVF International Conference on Computer Vision. New York City: IEEE; 2021. pp. 6824-6835

[179] Li Y, Wu CY, Fan H, Mangalam K, Xiong B, Malik J, et al. MVITV2: Improved multiscale vision transformers for classification and detection. In: Proceedings of the IEEE/CVF Conference on Computer Vision and Pattern Recognition. New York City: IEEE; 2022. pp. 4804-4814

[180] Ali A, Touvron H, Caron M, Bojanowski P, Douze M, Joulin A, et al. XCIT: Cross-covariance image transformers. Advances in Neural Information Processing Systems. 2021;**34**:20014-20027

[181] Tian Y, Ye Q, Doermann D. YOLOv12: Attention-centric real-time object detectors. arXiv preprint. 2025;**1**:1-13. arXiv preprint arXiv:2502.12524

[182] Everingham M, Van Gool L, Williams CK, Winn J, Zisserman A. The Pascal visual object classes (VOC) challenge. International Journal of Computer Vision. 2010;**88**:303-338

[183] Russakovsky O, Deng J, Su H, Krause J, Satheesh S, Ma S, et al. Imagenet large scale visual recognition challenge. International Journal of Computer Vision. 2015;**115**:211-252

[184] Lin TY, Maire M, Belongie S, Hays J, Perona P, Ramanan D, et al. Microsoft COCO: Common objects in context. In: Computer Vision–ECCV 2014: 13th European Conference, Zurich, Switzerland, September 6-12, 2014, Proceedings, Part v 13; Zurich, Switzerland. Cham, Switzerland: Springer International Publishing; 2014. pp. 740-755

[185] Krasin I, Duerig T, Alldrin N, Ferrari V, Abu-El-Haija S, Kuznetsova A, et al. OpenImages: A public dataset for large-scale multi-label and multi-class image classification. 2017. Available from https://github.com/openimages

[186] Shao S, Li Z, Zhang T, Peng C, Yu G, Zhang X, et al. Objects365: A large-scale, high-quality dataset for object detection. In: Proceedings of the IEEE/CVF International Conference on Computer

Vision. New York City: IEEE; 2019. pp. 8430-8439

[187] Kirillov A, Mintun E, Ravi N, Mao H, Rolland C, Gustafson L, et al. Segment anything. In: Proceedings of the IEEE/CVF International Conference on Computer Vision. New York City: IEEE; 2023. pp. 4015-4026. Available from: https://segment-anything.com

[188] Everingham M. The Pascal Visual Object Classes Challenge (VOC 2007) Results. Heidelberg, Germany: Springer Science+Business Media; 2007. Available from: http://pascallin.ecs.soton.ac.uk/challenges/VOC/voc2007/index.html

[189] Everingham M, Van Gool L, Williams CKI, Winn J, Zisserman A. The Pascal Visual Object Classes Challenge 2012 (VOC2012) Results. Heidelberg, Germany: Springer Science+Business Media; 2012. Available from: http://www.pascal-network.org/challenges/VOC/voc2012/workshop/index.html

[190] Deng J, Dong W, Socher R, Li LJ, Li K, Fei-Fei L. ImageNet: A large-scale hierarchical image database. In: 2009 IEEE Conference on Computer Vision and Pattern Recognition; Miami, Florida, USA. New York City: IEEE; 2009. pp. 248-255

[191] Oksuz K, Cam BC, Akbas E, Kalkan S. Localization recall precision (LRP): A new performance metric for object detection. In: Proceedings of the European Conference on Computer Vision (ECCV). Heidelberg, Germany: Springer Science+Business Media; 2018. pp. 504-519

[192] Zoph B, Le QV. Neural architecture search with reinforcement learning. arXiv preprint. 2016;**2**:1-16. arXiv preprint arXiv:1611.01578

[193] Chen Y, Yang T, Zhang X, Meng G, Pan C, Sun J. DetNAS: Backbone search for object detection. arXiv preprint. 2019;**4**:1-12. arXiv:1903.10979v4

[194] Liang T, Wang Y, Tang Z, Hu G, Ling H. OPANAS: One-shot path aggregation network architecture search for object detection. In: Proceedings of the IEEE/CVF Conference on Computer Vision and Pattern Recognition. New York City: IEEE; 2021. pp. 10195-10203

[195] Ghiasi G, Lin TY, Le QV. NAS-FPN: Learning scalable feature pyramid architecture for object detection. In: Proceedings of the IEEE/CVF Conference on Computer Vision and Pattern Recognition. New York City: IEEE; 2019. pp. 7036-7045

[196] Guo J, Han K, Wang Y, Zhang C, Yang Z, Wu H, et al. Hit-detector: Hierarchical trinity architecture search for object detection. In: Proceedings of the IEEE/CVF Conference on Computer Vision and Pattern Recognition. New York City: IEEE; 2020. pp. 11405-11414

[197] Xu H, Yao L, Zhang W, Liang X, Li Z. Auto-FPN: Automatic network architecture adaptation for object detection beyond classification. In: Proceedings of the IEEE/CVF International Conference on Computer Vision. New York City: IEEE; 2019. pp. 6649-6658

[198] Yao L, Xu H, Zhang W, Liang X, Li Z. SM-NAS: Structural-to-modular neural architecture search for object detection. In: Proceedings of the AAAI Conference on Artificial Intelligence. Vol. 34, No. 07. Palo Alto, California USA: Association for the Advancement of Artificial Intelligence (AAAI) Press; 2020. pp. 12661-12668

[199] Wang N, Gao Y, Chen H, Wang P, Tian Z, Shen C, et al. NAS-FCOS: Fast neural architecture search for object detection. In: Proceedings of the IEEE/

CVF Conference on Computer Vision and Pattern Recognition. New York City: IEEE; 2020. pp. 11943-11951

[200] Du X, Lin TY, Jin P, Ghiasi G, Tan M, Cui Y, et al. SPINENET: Learning scale-permuted backbone for recognition and localization. In: Proceedings of the IEEE/CVF Conference on Computer Vision and Pattern Recognition. New York City: IEEE; 2020. pp. 11592-11601

[201] Wu B, Li C, Zhang H, Dai X, Zhang P, Yu M, et al. FBNETV5: Neural architecture search for multiple tasks in one run. arXiv preprint. 2021;**3**. arXiv preprint arXiv:2111.10007:1-16

[202] Xiong Y, Liu H, Gupta S, Akin B, Bender G, Wang Y, et al. MobileDets: Searching for object detection architectures for mobile accelerators. In: Proceedings of the IEEE/CVF Conference on Computer Vision and Pattern Recognition. New York City: IEEE; 2021. pp. 3825-3834

[203] Goodfellow I, Pouget-Abadie J, Mirza M, Xu B, Warde-Farley D, Ozair S, et al. Generative adversarial networks. Communications of the ACM. 2020;**63**(11):139-144

[204] Wang TC, Liu MY, Zhu JY, Tao A, Kautz J, Catanzaro B. High-resolution image synthesis and semantic manipulation with conditional GANS. In: Proceedings of the IEEE Conference on Computer Vision and Pattern Recognition. New York City: IEEE; 2018. pp. 8798-8807

[205] Gulrajani I, Ahmed F, Arjovsky M, Dumoulin V, Courville AC. Improved training of Wasserstein GANS. Advances in Neural Information Processing Systems. 2017;**30**:1-11

[206] Zhu JY, Park T, Isola P, Efros AA. Unpaired image-to-image translation using cycle-consistent adversarial networks. In: Proceedings of the IEEE International Conference on Computer Vision. New York City: IEEE; 2017. pp. 2223-2232

[207] Radford A, Kim JW, Hallacy C, Ramesh A, Goh G, Agarwal S, et al. Learning transferable visual models from natural language supervision. In: International Conference on Machine Learning; Virtual Conference. Cambridge, MA, USA: PMLR; 2021. pp. 8748-8763

[208] Jia C, Yang Y, Xia Y, Chen YT, Parekh Z, Pham H, et al. Scaling up visual and vision-language representation learning with noisy text supervision. In: International Conference on Machine Learning; Virtual Conference. Cambridge, MA, USA: PMLR; 2021. pp. 4904-4916

[209] Li LH, Yatskar M, Yin D, Hsieh CJ, Chang KW. VisualBERT: A simple and performant baseline for vision and language. arXiv preprint. 2019;**1**:1-14. arXiv preprint arXiv:1908.03557

[210] Marcus G, Davis E, Aaronson S. A very preliminary analysis of DALL-E 2. arXiv preprint. 2022;**2**:1-14. arXiv preprint arXiv:2204.13807

[211] Singh A, Hu R, Goswami V, Couairon G, Galuba W, Rohrbach M, et al. FLAVA: A foundational language and vision alignment model. In: Proceedings of the IEEE/CVF Conference on Computer Vision and Pattern Recognition. 2022. pp. 15638-15650

[212] Li Y, Li J, Lin W, Li J. Tiny-DSOD: Lightweight object detection for resource-restricted usages. arXiv preprint. 2018;**1**:1-12. arXiv preprint arXiv:1807.11013

[213] Lyu R. NanoDet-Plus: Super Fast and High Accuracy Lightweight

Anchor-Free Object Detection Model. Maryland, USA: Apache Software Foundation; 2021. Available from: https://github.com/RangiLyu/nanodet

[214] Cai Y, Li H, Yuan G, Niu W, Li Y, Tang X, et al. YOLOBILE: Real-time object detection on mobile devices via compression-compilation co-design. In: Proceedings of the AAAI Conference on Artificial Intelligence. Vol. 35, No. 2. Palo Alto, California USA: Association for the Advancement of Artificial Intelligence (AAAI) Press; 2021. pp. 955-963

[215] Tan M, Le Q. EfficientNet: Rethinking model scaling for convolutional neural networks. In: International Conference on Machine Learning; California, USA. Cambridge, MA, USA: PMLR; 2019. pp. 6105-6114

Chapter 6

Usage of Wavelets in Image-Based Steganography

Davi Schmitz, Armando Leopoldo Keller, Rodrigo Marques de Figueiredo, Vitor Camargo Nardelli and Jean Schmith

Abstract

This work explores steganography as a tool for embedding hidden information, such as watermarks, into images. Encoding and decoding methods based on the Discrete Wavelet Transform (DWT) and Singular Value Decomposition (SVD) were proposed. The study provides a comprehensive comparison of the steganographic behavior of models using low-frequency, high-frequency, and multi-level sub-bands, evaluating the impact of different wavelet families, scaling factors, and decomposition levels. Metrics such as PSNR and MSE were used to assess the results, which were compared with the existing literature. The experiments demonstrated that high-frequency techniques offer the best balance between imperceptibility and extraction quality, achieving PSNR values of up to 51.19 dB when using the sym16 wavelet, while the multilevel model presented higher security at the cost of lower decoding quality. These findings help address the unexplored gaps in the literature, providing a detailed analysis of the possibilities and limitations of these techniques.

Keywords: steganography, digital security, image processing, wavelet transform, single value decomposition

1. Introduction

In an increasingly digital world, where vast amounts of data can be transmitted effortlessly, concerns about data integrity and the authenticity of information have grown substantially. Ensuring that the information consumed is reliable and unchanged has become a critical challenge. In response, the fields of cryptography and steganography have become prominent focuses of research and practical application.

Cryptography focuses on protecting the content of messages or media by transforming it into an unreadable format for unauthorized parties. In contrast, steganography revolves around concealing the very existence of a message, embedding it within another medium in such a way that its presence is undetectable to unintentional observers [1]. Although distinct, these disciplines are complementary to address modern security and privacy concerns.

Focusing specifically on image data, steganography enables the embedding of hidden information into digital images without significantly altering their appearance.

This capability has paved the way for applications such as digital watermarking, where unique identifiers can be embedded in images. These identifiers serve to verify ownership, attest to the authenticity of the images, and ensure that they have not been tampered with [2]. Digital watermarking, as a subset of steganography, provides a robust mechanism for protecting intellectual property and maintaining trust in digital assets.

In the current literature, steganography is primarily implemented using two approaches: information hiding in the spatial domain and the frequency domain. In the spatial domain, the most common technique is the Least Significant Bit (LSB) method, which embeds hidden data by modifying the least significant bits of the cover image. Although this technique is straightforward and computationally efficient, it is more vulnerable to detection and distortion through methods such as histogram analysis [3] and RS steganalysis [4]. In contrast, the frequency domain achieves steganography by leveraging transformations such as the Discrete Cosine Transform (DCT) and Discrete Wavelet Transform (DWT), which provide greater robustness and imperceptibility by embedding data into transformed coefficients rather than directly altering pixel values.

Focusing specifically on previous research in image-based steganography using DWT, **Table 1** has been compiled to summarize these contributions.

Based on the studies analyzed in the literature review, it is possible to observe that even when considering only approaches that use the Discrete Wavelet Transform to perform steganography in images, there is significant variation in the choice of wavelet families, decomposition levels, and DWT coefficients used for embedding. With respect to the chosen wavelets mother, the analyzed studies do not provide comparisons of the results obtained with different wavelet families, nor with respect to the decomposition level. For example, studies such as Prabakaran and Bhavani [7] and Hussain et al. [9] use a two-level decomposition, but do not present metrics comparing the results obtained with other levels, which presents an opportunity for the present work.

Article	Objective	Wavelets	DWT Level	Sub-bands
[5]	Steganography in Images Using Redundant DWT (RDWT) and QR Factorization.	Not informed	1	HH
[6]	Steganography in Images Using DWT and SVD: Applying Watermarks to Specific Frames of Video Files.	Haar	1	LH
[7]	Steganography in images using DWT and Alpha blending.	Haar	2	LH, HL, HH
[8]	Steganography in grayscale images using DWT and SVD.	Not informed	1	HH
[9]	Steganography in images using DWT and SVD. The embedding occurs in frames of a video file.	Haar	2	LL, LH, HL, HH
[10]	Steganography in grayscale images using DWT and SVD.	Not informed	2	HL
[11]	Steganography in images using DWT and SVD.	Daubechies	1	LL, LH, HL, HH

Table 1.
Summary of related works.

This study aims to implement three methods of DWT image steganography, low frequency, high frequency, and multilevel embedding, comparing their results in terms of the Peak Signal-to-Noise Ratio (PSNR) for a wide range of wavelet families, scaling factors, and decomposition levels, something not found in the current literature.

1.1 Discrete wavelet transform - DWT

In this work, the main mathematical instrument used to achieve steganography is the Discrete Wavelet Transform (DWT). DWT is widely applied in image processing due to its ability to capture both global and local characteristics of an image [12, 13]. The technique involves decomposing the image into sub-bands of different frequencies through the application of low-pass and high-pass filters regarding the wavelet mother function. This decomposition results in varying levels of detail, ranging from low-frequency components, which represent the global approximation of the image, to high-frequency components, which emphasize edges and fine details [14].

In two-dimensional applications such as images, DWT decomposition is performed first along the vertical direction and then along the horizontal direction. This process produces sub-bands such as Low-Low (LL), Low-High (LH), High-Low (HL), and High-High (HH), representing the approximation, vertical details, horizontal details, and diagonal details, respectively. These sub-bands are generated by applying a combination of low-pass and high-pass filters to the rows and columns of the image. It is also possible to achieve multiple decomposition levels by using the LL sub-band from the previous level as the input for the next. This iterative process allows for a hierarchical representation of the image, capturing its structure at multiple resolutions [15].

As each decomposition level uses the previous LL sub-band as input, it also halves the size of the coefficients. For instance, in a 512x512-pixel image, the first DWT level results in an approximation sub-band of 256x256 pixels. Applying the DWT again to this sub-band produces a new LL band with a size of 128x128 pixels. **Figure 1** displays the resulted sub-bands of a three-level DWT decomposition.

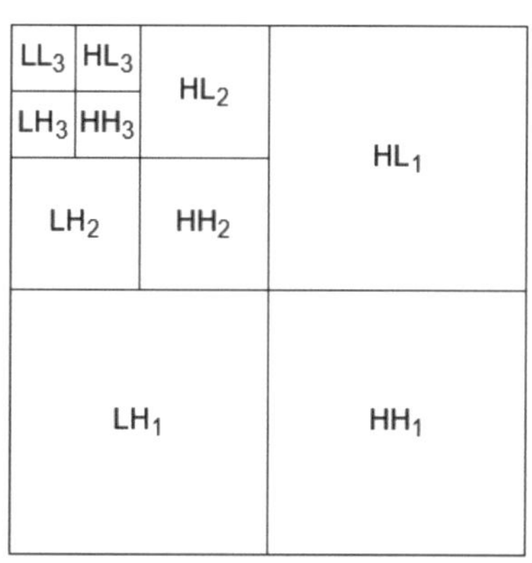

Figure 1.
Example of three level DWT decomposition [15].

2. Methodology

This section describes the implementation process for the different steganography methods compared in this study. These methods were implemented using the Python programming language, the OpenCV [16], and PyWavelet [17] libraries and are available in Github; please use the QR Code in the examples to access it.

2.1 Images

To ensure the standardization of the images used for algorithm testing, the USC-SIPI Image Database [18] was used, a widely recognized resource in image processing research. This database contains images of varying sizes, resolutions, and characteristics, enabling the selection of appropriate images to test various aspects of steganography algorithms. All selected images are in TIFF format, known for its larger file sizes due to lossless compression. The primary test image chosen was "peppers," a 512x512 pixel color image frequently used as a benchmark in image processing studies due to its visual complexity, which includes diverse textures, edges, and color variations.

In addition to the Mean Squared Error (MSE) and Peak Signal-to-Noise Ratio (PSNR) metrics, which are used to quantify the similarity between the cover image and the stego image, a QR Code with a resolution of 512x512 pixels in TIFF format was chosen as the embedded image. The successful reading of this QR Code after the decoding process will indicate a successful data extraction. This QR Code leads to the Github link with the implementation of this study.

2.2 Encoding process

This section outlines the common steps involved in the steganographic encoding process used in all implemented methods. Specific details of the different frequency embeddings can be found in subsection 2.4, subsection 2.5, and subsection 2.6.

First, the cover and embedded images are loaded and split into their respective RGB color channels. The Discrete Wavelet Transform (DWT) is then applied to each RGB channel, extracting the sub-bands discussed in subsection 1.1.

Once the DWT coefficients are obtained, the Singular Value Decomposition (SVD) is calculated for the relevant sub-bands, depending on the specific embedding method being used. SVD is a matrix decomposition technique commonly used in image processing to provide a systematic way to derive a low-dimensional approximation of high-dimensional data in terms of dominant patterns [19]. When SVD is applied to an image, it produces three stable matrices that encapsulate the primary characteristics of the image.

For any given matrix A of size $m \times n$, the SVD factorization is given by Eq. (1), resulting in three matrices, where $\mathcal{U}$ is an orthogonal matrix of $m \times m$, V is an orthogonal matrix of size $n \times n$, and Σ is a diagonal matrix (containing singular values), which represents the magnitudes of the vectors ($\mathcal{U}$) and (V^T), which capture the intrinsic geometric structure and important features of the original matrix A.

In the context of steganography, modifying the singular value matrix, rather than directly altering the DWT coefficients, allows for a more seamless embedding with minimal visual impact.

$$A = \mathcal{U}\Sigma V^T \tag{1}$$

Having the SVD diagonal matrix for the desired DWT coefficients, a new diagonal matrix is computed using Eq. (2), where Σ_s is the stego calculated matrix, Σ_c is the diagonal matrix of the cover image, Σ_e is the diagonal matrix of the embedded image, and α is the scaling factor.

$$\Sigma_s = \Sigma_c + (\Sigma_e \alpha) \tag{2}$$

The scaling factor α determines the intensity of modifications applied to the singular values of the cover image, effectively controlling the influence of the embedded image on it. A larger scaling factor indicates that a greater portion of the singular values of the embedded image are incorporated, resulting in a higher quality of the image extracted during the decoding process. However, this also leads to more noticeable visual changes in the stego image. Conversely, a smaller scaling factor reduces each singular value of the embedded image before incorporation, minimizing perceptible alterations in the stego image but yielding lower quality in the extracted image during decoding. In this study, the following scaling factors were used: 0.02, 0.05, 0.1, and 0.2.

Finally, the modified diagonal matrices are combined with the unmodified matrices $\mathcal{U}$ and V^T of the cover image to reconstruct the DWT coefficients, now containing hidden information. In the final step, the Inverse Discrete Wavelet Transform (IDWT) is applied to recover the color channels, which are then combined to form the steganographic image. The steps that compose the encoding process are depicted in **Figure 2**.

2.3 Decoding process

This section outlines the common steps involved in the steganographic decoding process used in all the methods implemented here. Specific details of the different frequency embeddings can be found in subsection 2.4, subsection 2.5, and subsection 2.6.

In the decoding process, the stego image is first loaded and split into its RGB color channels. The Discrete Wavelet Transform (DWT) coefficients are then extracted for each channel, mirroring the steps in the encoding process. It is necessary to know what

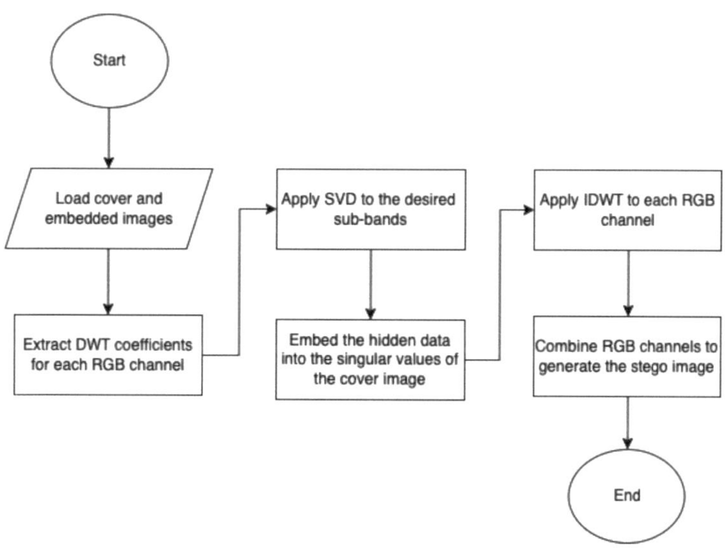

Figure 2.
Flowchart of the encoding process.

Wavelet mother was used for the encoding, as the same one should be used for decoding. Subsequently, the Singular Value Decomposition (SVD) is applied to the relevant sub-bands, depending on the specific encoding method that was used. Having the diagonal matrix Σ_s of the stego image and the scaling factor α used during the encoding, it is possible to retrieve the diagonal matrix of the embedded image Σ_e using Eq. (3).

$$\Sigma_e = \frac{\Sigma_s}{\alpha} \tag{3}$$

Once the diagonal matrix Σ_e is found, it can be multiplied by the original matrices $\mathcal{U}$ and V^T of the embedded image for each RGB channel. This process reconstructs the DWT coefficients of the sub-band used for encoding. Finally, the Inverse Discrete Wavelet Transform (IDWT) is applied to these reconstructed coefficients, and the RGB channels are combined, resulting in the extracted hidden image. The steps that compose the decoding process are shown in **Figure 3**.

2.4 Low-frequency embedding

The first steganographic method implemented involved embedding the information from the hidden image into the LL sub-band of the cover. This sub-band is derived by applying low-pass filters in both the horizontal and vertical directions, resulting in an approximation of the image that retains its main features. Although this method has a simpler implementation, dealing with a single sub-band, as the LL sub-band encapsulates the main structure of the image, even minor changes to its coefficients become visually noticeable, leading to a decrease in the PSNR values when comparing the cover with the steganographic image.

Figure 4 presents, in the first row, the cover and embedded images used for testing the low-frequency method. The second and third rows depict the sub-bands

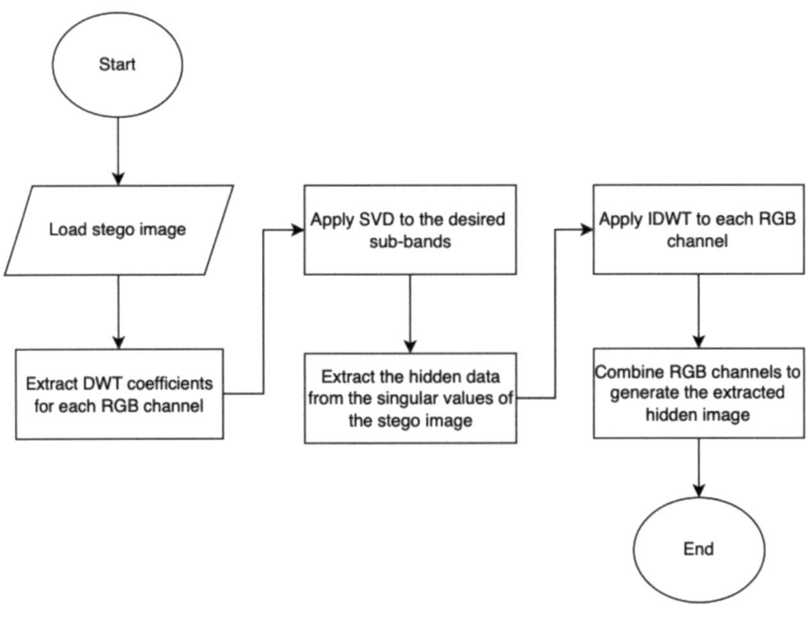

Figure 3.
Flowchart of the decoding process.

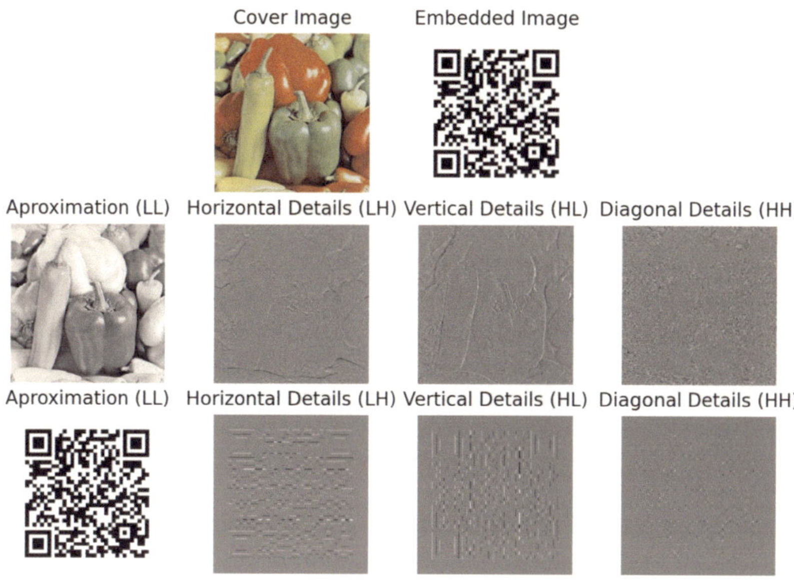

Figure 4.
Sub-bands produced by the DWT.

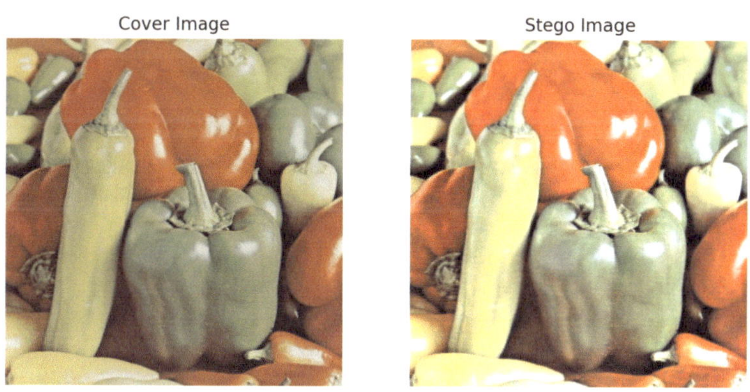

Figure 5.
Result of the low-frequency steganography.

obtained by applying the Haar DWT to both images. In this method, only the LL approximation is used.

Figure 5 illustrates the comparison between the cover and the stego image, generated by the low-frequency method and containing the QR Code shown in **Figure 4** hidden inside. In this example, the Haar wavelet was used with a scaling factor of 0.2. The comparison reveals significant visual differences between the images, indicating poor performance of the method in terms of imperceptibility.

2.5 High-frequency embedding

To overcome the limitations of embedding data in the LL sub-band of the DWT, the high-frequency steganography method embeds only in the HL, LH, and

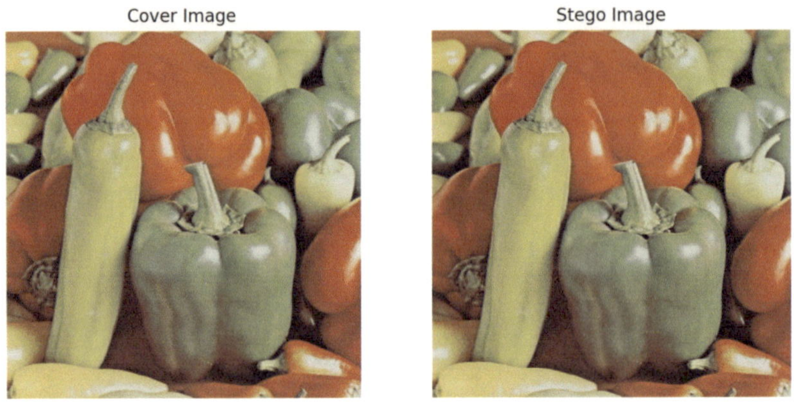

Figure 6.
Result of the high-frequency steganography.

HH coefficients. As presented in subsection 1.1, these sub-bands use a combination of low-pass and high-pass filters to capture the horizontal, vertical, and diagonal details of the images. These coefficients are referred to as high frequency because they capture rapid variations in image intensity, such as abrupt changes, edges, and fine details. Since these details do not significantly affect the basic structure of the image, embedding data in them allows for better imperceptibility, at the cost of increased implementation complexity due to the presence of three distinct bands. The encoding process embeds the singular values of the HL, LH, and HH sub-band coefficients from the hidden image into the corresponding singular values of the HL, LH, and HH sub-band coefficients of the cover image for each color channel.

For the same example presented in subsection 2.4, of applying the Haar wavelet and a scaling factor of 0.2, the result of the high-frequency embedding can be seen in **Figure 6**. A significant improvement in visual imperceptibility can be observed with this embedding method compared to the result obtained for low frequencies, which also translates to better PSNR results, since this metric provides a quantitative measure of the difference between the cover and stego images.

2.6 High-frequency embedding with multilevel DWT

As explained in subsection 1.1, after obtaining the DWT coefficients LL, LH, HL, and HH, the approximation sub-band LL can be further decomposed, generating new coefficients LL2, LH2, HL2, and HH2. Multilevel embedding involves using the coefficients of the higher decomposition level to hide the data of the embedded image. For instance, in the case of a three-level DWT, the embedding would happen only in the LH3, HL3, and HH3 sub-bands. This method provides greater security because, in addition to knowing the wavelet used and the scaling factor, successful data extraction also requires knowing the specific decomposition level where the data was embedded.

However, the usage of higher levels of decomposition comes with a penalty in the imperceptibility of the method. As described in subsection 1.1, at each decomposition level, the size of the coefficients is halved, resulting in fewer bits available for data embedding.

Figures 7 and **8** display the coefficients obtained from a 2-level decomposition, along with the diagram that illustrates the naming convention for each coefficient.

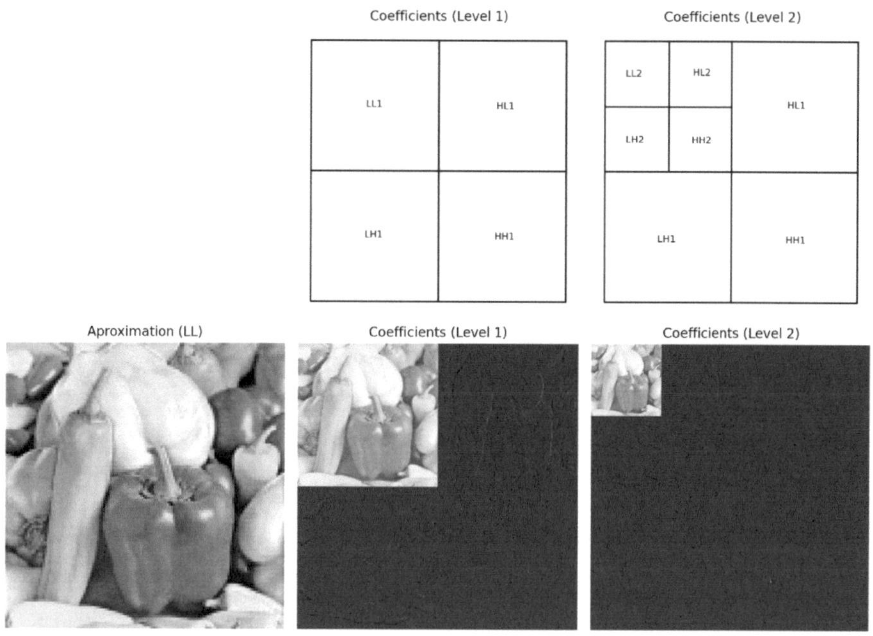

Figure 7.
Coefficients obtained by applying Multilevel DWT to the cover image.

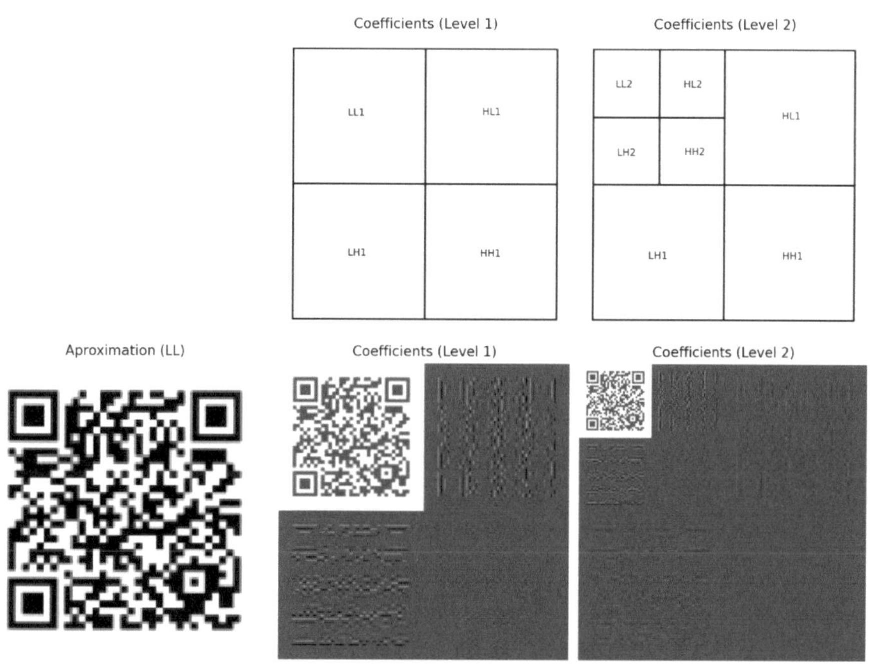

Figure 8.
Coefficients obtained by applying Multilevel DWT to the embedded image.

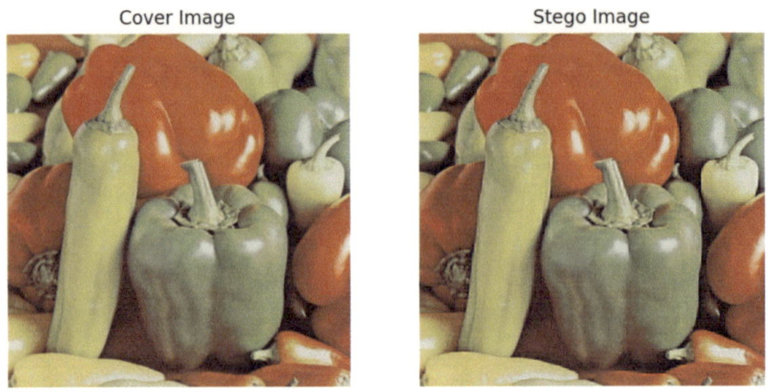

Figure 9.
Result of the Multilevel DWT steganography.

Parameter	Tested Values
Wavelet mother	Haar, db1, db8, db16, sym2, sym8, sym16
Scale factor α	0.02, 0.05, 0.1, 0.2
DWT level for encoding	1, 2, 3, 4
DWT level for decoding	1, 2, 3, 4

Table 2.
Benchmark parameters.

For comparison purposes with the other implemented methods, the embedding illustrated in **Figure 9** is done with the Haar wavelet, a scaling factor of 0.2, and a 2-level DWT decomposition. At this level of decomposition, the method provides an imperceptibility similar to the high frequency presented in subsection 2.5.

2.7 Metrics

For the collection of metrics, a benchmark was developed to execute the steganographic models in a parameterized manner, altering aspects such as the wavelet mother used, the scaling factor, and the decomposition level of the wavelet transform. All parameters and tested values are listed in **Table 2**.

3. Results

This section discusses the collected PSNR metrics for the low-frequency, high-frequency and multilevel methods. The Peak Signal-to-Noise Ratio (PSNR) measures the distortion between the cover image and the stego image. A higher PSNR value indicates a better imperceptibility of the steganographic method, which makes embedded information less noticeable. PSNR values greater than 40 dB are considered excellent and reflect minimal visual distortion. Values between 30 and 40 dB are deemed reasonable, with some noticeable but acceptable distortion. PSNR values below 30 dB are generally unacceptable, indicating significant distortion that degrades the visual quality of the stego image.

3.1 Low frequency

In the low-frequency embedding method, the PSNR results were generally low across various wavelet mothers and scaling factors used. The best results were achieved with the Haar and db1 wavelets when using the smallest scaling factor tested of 0.02, reaching a PSNR of 36.87 dB. This indicates that the method is too sensitive to the data embedding, which demands the usage of very low scaling factors, which impacts the decoding performance.

Figure 10 displays the decoding result of the method when using a Haar wavelet and a scaling factor of 0.02. The embedded QR Code can be successfully decoded, but the image shows visible alterations. The complete PSNR results can be found in **Figure 11**. Note that all wavelets mothers showed the same behavior.

3.2 High frequency

In the high-frequency embedding method, there is an expressive improvement in the PSNR results compared to the low-frequency ones. The best results were obtained with the sym16 wavelet, ranging from 41 to 51.19 dB when testing with scaling factors of 0.2 to 0.02, respectively. These results indicate a better imperceptibility of the

Figure 10.
Decoding result for the low-frequency embedding.

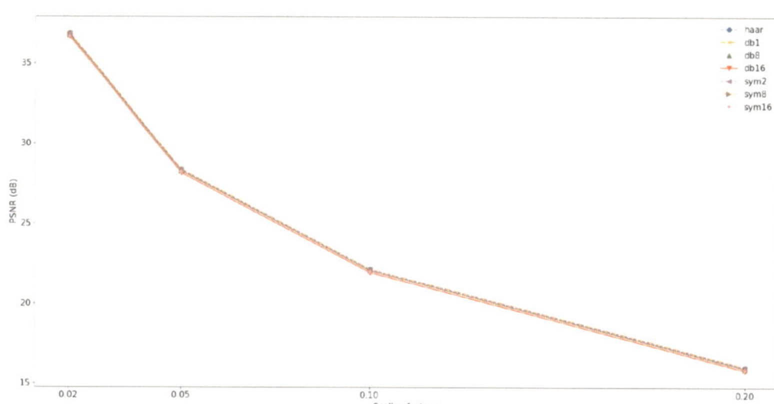

Figure 11.
PSNR results for the low-frequency embedding.

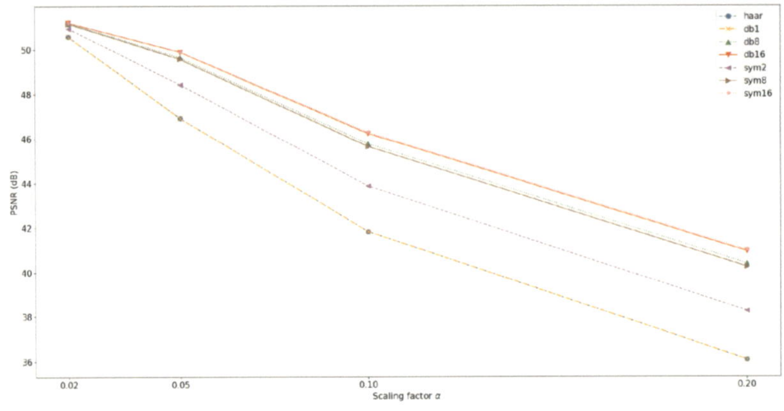

Figure 12.
PSNR results for the high-frequency embedding.

method, which alters only fine details while keeping the basic structure of the cover image unchanged. The complete PSNR results can be found in **Figure 12**. The worst results were with Haar and Daubechies of the first order.

3.3 High-frequency embedding with multilevel DWT

In the high-frequency embedding with multilevel DWT, the PSNR results were slightly inferior to the ones found when encoding at a single level, but still superior to the low-frequency embedding. The best result was with the sym16 wavelet, ranging from 37.7 to 50.83 dB using scaling factors of 0.2 and 0.02, respectively. The complete PSNR results can be found in **Figure 13**.

Regarding the effect of different decomposition levels, it was observed that with each additional level, there is a decrease in the PSNR results between the cover and stego images. This effect is illustrated in **Figure 14**, which shows the results using the Haar wavelet and a scaling factor of 0.2, with variations only in the decomposition level.

Furthermore, increasing the decomposition level also significantly reduces the quality of the decoded image, as demonstrated by the comparison between **Figures 15**

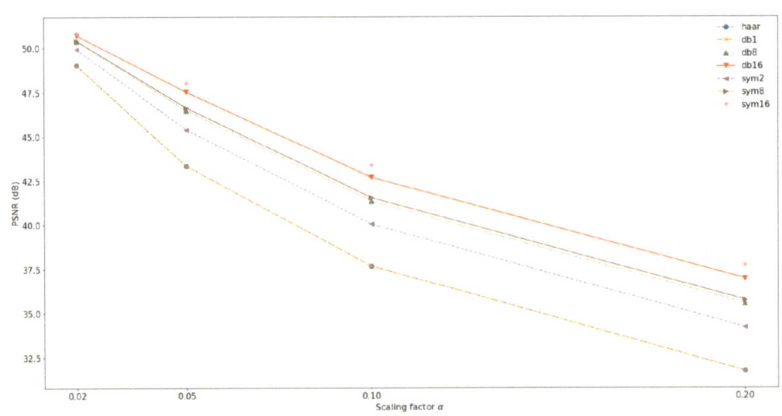

Figure 13.
PSNR Results for the multilevel embedding.

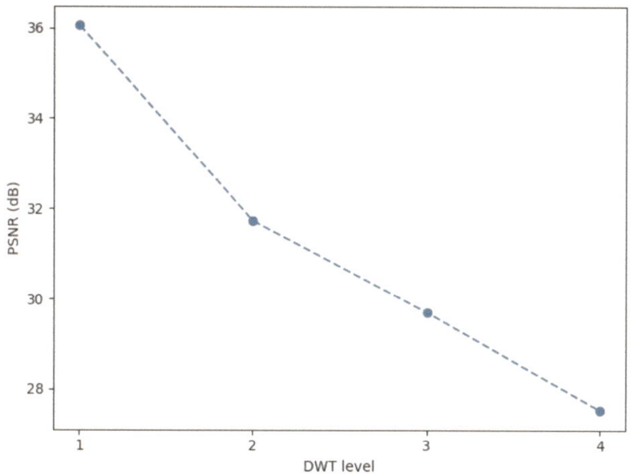

Figure 14.
PSNR results for different decomposition levels.

Figure 15.
Decoding performance with DWT level 2.

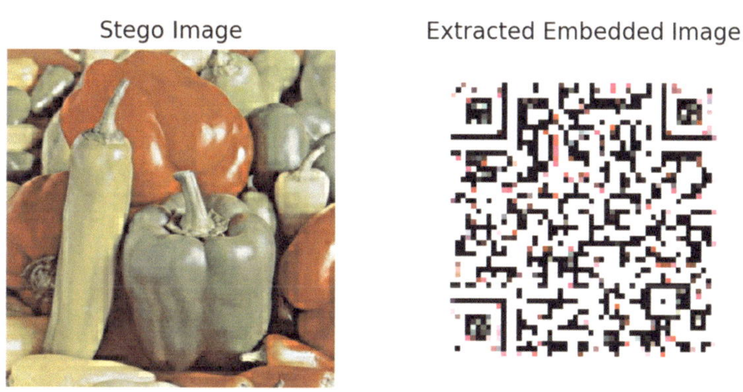

Figure 16.
Decoding performance with DWT level 4.

and **16**. In the latter case, the degradation was so severe that the QR Code became unreadable.

Thus, the choice of the decomposition level to be used should balance the increased security provided by multilevel embedding, since this additional information becomes essential for decoding, with the impact that higher levels can have on the imperceptibility of the steganographic method and the quality of the decoded image.

4. Discussion

The study provides a detailed comparison of different embedding methods in DWT-based steganography systems. This analysis, previously unexplored in the literature, emphasizes the clear benefits of utilizing the higher-frequency detail coefficients of the DWT. Although this method introduces a slight increase in implementation complexity, it successfully embeds hidden data without altering the LL sub-band, preserving the most structural and visually significant parts of the image, while consistently delivering superior performance across all tested wavelet families.

Comparing our results with those presented in **Table 1**, we presented a more complete exploration on the choice of wavelet mother. Note that most of the works choose the Haar family and one uses Daubechies. One might note in **Figure 12** that Haar and Daubechies of the first-order wavelet mother had the worst result considering the high-frequency embedding.

With regard to multilevel wavelet decomposition, the findings reveal the need for a trade-off. Higher decomposition levels can enhance security, as the level information becomes essential for decoding; however, they also negatively affect the quality of both the stego and decoded images. Specifically, our tests revealed that decomposition levels above 2 led to a significant drop in PSNR values while using a 0.2 scaling factor, falling below 30 dB, a threshold commonly associated with low image quality across all wavelet families as can be seen in the examples of **Figures 13** and **14**. This highlights the balance required between increased security and maintaining image quality.

The experimental results revealed significant differences between the models. Low-frequency embedding achieved reasonable results only with very low scaling factors (e.g., 0.02), since altering the LL sub-band, where the primary structural features of the cover image are concentrated, caused noticeable visual degradation at higher scaling factors. Despite this limitation, the extracted hidden image retained adequate quality, with the QR Code remaining readable. High-frequency embedding improved imperceptibility by altering only fine details in the LH, HL, and HH sub-bands, achieving satisfactory PSNR values even with higher scaling factors (e.g., 0.2). The best performance was observed with the Sym16 wavelet, which reached 51.19 dB at a scaling factor of 0.02.

5. Conclusion

This work aimed to explore and compare different steganographic techniques applied to images using Discrete Wavelet Transform (DWT) and Singular Value Decomposition (SVD). The primary goal was to implement methods for embedding one image into another while evaluating low-frequency, high-frequency, and multilevel embedding approaches, providing a wide comparison in terms of Wavelet family, scaling factors, and decomposition levels. The study included a theoretical analysis of steganography, DWT, and SVD, followed by experiments testing the various proposed configurations. Metrics such as PSNR and visual analysis were used to assess imperceptibility and extraction quality, with a QR Code as the hidden image to validate readability after decoding. Multilevel decomposition improved security by requiring information from the decomposition level for decoding but reduced PSNR and extraction quality at higher levels, with the best results comparable to single-level high-frequency methods.

In conclusion, this study demonstrated the effectiveness of DWT and SVD for image steganography, offering a comprehensive analysis of embedding techniques in various configurations. These findings can serve as input for the decision-making process in the design and optimization of steganographic systems.

Acknowledgements

This work has partially funded and supported by the Hardware Competence Center for Digital Agriculture, with financial resources from PPI HardwareBR of the MCTI grant number 056/2023, signed with EMBRAPII.

Conflict of interest

The authors declare no conflict of interest.

Author details

Davi Schmitz[1†], Armando Leopoldo Keller[1†], Rodrigo Marques de Figueiredo[1,2†], Vitor Camargo Nardelli[2†] and Jean Schmith[1,2*†]

1 Unisinos University, São Leopoldo, Brazil

2 Competence Center on Digital Agriculture (EMBRAPII), SENAI Innovation Institute for Sensor Systems (ISI-SIM), São Leopoldo, Brazil

*Address all correspondence to: jean.schmith@senairs.org.br; j.schmith@gmail.com

† These authors contributed equally.

References

[1] Petitcolas FAP, Anderson RJ, Kuhn MG. Information hiding—A survey. Proceedings of the IEEE. 1999; **87**(7):1062-1078

[2] Kundur D, Hatzinakos D. A robust digital image watermarking method using wavelet-based fusion. In: Proceedings of International Conference on Image Processing. Vol. 1. Santa Barbara, USA: IEEE; 1997. pp. 544-547

[3] Zhang T, Ping X. A new approach to reliable detection of LSB steganography in natural images. Signal Processing. 2003;**83**(10):2085-2093

[4] Fridrich J, Goljan M, Du R. Reliable detection of LSB steganography in color and grayscale images. In: Proceedings of the 2001 Workshop on Multimedia and Security: New Challenges. Ottawa, Canada: Association for Computing Machinery; 2001. pp. 27-30

[5] Subhedar MS, Mankar VH. Image steganography using redundant discrete wavelet transform and QR factorization. Computers & Electrical Engineering. 2016;**54**:406-422

[6] Liu Q, Yang S, Liu J, Xiong P, Zhou M. A discrete wavelet transform and singular value decomposition-based digital video watermark method. Applied Mathematical Modelling. 2020; **85**:273-293

[7] Prabakaran G, Bhavani R. A modified secure digital image steganography based on discrete wavelet transform. In: 2012 International Conference on Computing, Electronics and Electrical Technologies (ICCEET). Nagercoil, India: IEEE; 2012. pp. 1096-1100

[8] Subhedar MS, Mankar VH. High capacity image steganography based on discrete wavelet transform and singular value decomposition. In: Proceedings of the 2014 International Conference on Information and Communication Technology for Competitive Strategies. Udaipur, India: Association for Computing Machinery; 2014. pp. 1-7

[9] Hussain AA et al. Multi-level Steganography System Using Wavelet Transform. Journal of Engineering and Sustainable Development. 2018;**22**(3): 50-61

[10] Zaidan FK. Digital Image Steganography Scheme Based on DWT and SVD. Diyala Journal of Engineering Sciences. 2020;**13**(4):10-17

[11] Narasimmalou T, Allen JR. Optimized discrete wavelet transform based steganography. In: 2012 IEEE International Conference on Advanced Communication Control and Computing Technologies (ICACCCT). Ramanathapuram, India: IEEE; 2012. pp. 88-91

[12] Dias LS, Schmith J, de Oliveira JD, Copetti JB. Classification of two-phase flow pattern in microtubes by image processing and machine learning. International Journal of Multiphase Flow. 2025;**185**:105122

[13] Kinoshita NYK, Jean S, Augusto ME, de Figueiredo RM. A method for identifying vegetation under distribution power lines by remote sensing. Journal of Control, Automation and Electrical Systems. 2023;**34**(6):1284-1293

[14] Xu J, Sung AH, Shi P, Liu Q. JPEG compression immune steganography using wavelet transform. In: International Conference on Information Technology: Coding and Computing, 2004. Proceedings. ITCC 2004. Vol. 2. Las Vegas, USA: IEEE; 2004. pp. 704-708

[15] Parul M, Rohil H. Optimized image steganography using discrete wavelet transform (DWT). International Journal of Recent Development in Engineering and Technology. 2014;**2**(2):75-81

[16] Bradski G. The OpenCV library. Dr. Dobb's Journal of Software Tools. 2000

[17] Lee G, Gommers R, Waselewski F, Wohlfahrt K, O'Leary A. PyWavelets: A python package for wavelet analysis. Journal of Open Source Software. 2019; **4**(26):1237

[18] USC-SIPI Image Database. Available from: https://sipi.usc.edu/database/ [Accessed: January 25, 2025]

[19] Brunton S, Kutz N. Data-Driven Science and Engineering: Machine Learning, Dynamical Systems, and Control. Cambridge University Press; 2022